Split It Up:

More Fractions, Decimals, and Percents

STUDENT BOOK

TERC

Mary Jane Schmitt, Myriam Steinback,
Tricia Donovan, Martha Merson, and Marlene Kliman

Bothell, WA • Chicago, IL • Columbus, OH • New York, NY

TERC
2067 Massachusetts Avenue
Cambridge, Massachusetts 02140

EMPower Research and Development Team
Principal Investigator: Myriam Steinback
Co-Principal Investigator: Mary Jane Schmitt
Research Associate: Martha Merson
Curriculum Developer: Tricia Donovan

Contributing Authors
Donna Curry
Marlene Kliman

Technical Team
Graphic Designer and Project Assistant: Juania Ashley
Production and Design Coordinator: Valerie Martin
Copyeditor: Jill Pellarin

Evaluation Team
Brett Consulting Group:
Belle Brett
Marilyn Matzko

EMPower™ was developed at TERC in Cambridge, Massachusetts. This material is based upon work supported by the National Science Foundation under award number ESI-9911410 and by the Education Research Collaborative at TERC. Any opinions, findings, and conclusions or recommendations expressed in this publication are those of the authors and do not necessarily reflect the views of the National Science Foundation.

TERC is a not-for-profit education research and development organization dedicated to improving mathematics, science, and technology teaching and learning.

http://empower.terc.edu

Printed in the United States of America
1 2 3 4 5 6 7 8 9 QDB 15 14 13 12 11

ISBN 978-0-07662-091-3
MHID 0-07-662091-3

Contents

Introduction

Welcome to EMPower

Students using the EMPower books often find that EMPower's approach to mathematics is different from the approach found in other math books. For some students, it is new to talk about mathematics and to work on math in pairs or groups. The math in the EMPower books will help you connect the math you use in everyday life to the math you learn in your courses.

We asked some students what they thought about EMPower's approach. We thought we would share some of their thoughts with you to help you know what to expect.

> *"It's more hands-on."*
>
> *"More interesting."*
>
> *"I use it in my life."*
>
> *"We learn to work as a team."*
>
> *"Our answers come from each other… [then] we work it out ourselves."*
>
> *"Real-life examples like shopping and money are good."*
>
> *"The lessons are interesting."*
>
> *"I can help my children with their homework."*
>
> *"It makes my brain work."*
>
> *"Math is fun."*

EMPower's goal is to make you think and to give you puzzles you will want to solve. Work hard. Work smart. Think deeply. Ask why.

Using This Book

This book is organized by lessons. Each lesson has the same format.

- The first page explains the lesson and states the purpose of the activity. Look for a question to keep in mind as you work.
- The activity pages come next. You will work on the activities in class, sometimes with a partner or in a group.
- Look for shaded boxes with additional information and ideas to help you get started if you become stuck.

- Practice pages follow the activities. These practices will make sense to you after you have done the activity. The three types of practice pages are

  Practice: Provides another chance to see the math from the activity and to use new skills.

  Extension: Presents a challenge with a more difficult problem or a new but related math idea.

  Test Practice: Asks a number of multiple-choice questions and one open-ended question.

In the *Appendices* at the end of the book, there is space for you to keep track of what you have learned and to record your thoughts about how you can use the information.

- Use notes, definitions, and drawings to help you remember new words in *Vocabulary*, pages 125–126.
- Answer the *Reflections* questions after each lesson, pages 127–134.

Tips for Success

Where do I begin?

Many people do not know where to begin when they look at their math assignments. If this happens to you, first try to organize your information. Read the problem. Start a drawing to show the situation.

Much of this unit is about parts and wholes.

Ask yourself:

What makes up the whole group? What number is just part of the group?

Another part of getting organized is figuring out what skills are required.

Ask yourself:

What do I already know? What do I need to find out?

Write down what you already know.

I cannot do it. It seems too hard.

Make the numbers smaller or friendlier. Try to solve the same problem with the benchmark fraction 1/2.

Ask yourself:

Have I ever seen something like this before? What did I do then?

You can always look back at another lesson for ideas.

Am I done?

Don't walk away yet. Check your answers to make sure they make sense.

Ask yourself:

Did I answer the question?

Does the answer seem reasonable? Do the conclusions I am drawing seem logical?

Check your math with a calculator. Ask others whether your work makes sense to them.

Practice

Complete as many of the practice pages as you can to sharpen your skills.

Opening the Unit: Split It Up!

What are some ways to split things up?

Comparing the whole amount with some part or parts is an ongoing theme in this unit. We use **percents**, **decimals**, and **fractions** to keep track of the relationships between **parts** and **wholes**.

In this session, you will explain some percents using words. You will look at percents and fractions on a set of coupons. Using what you know about percents, fractions, and decimals, you will put the coupons in order to show which coupons offer the biggest savings.

Activity 1: Making a Mind Map

Make a Mind Map using words, number, pictures, or ideas that come to mind when you think of *fractions*, *decimals*, and *percents*.

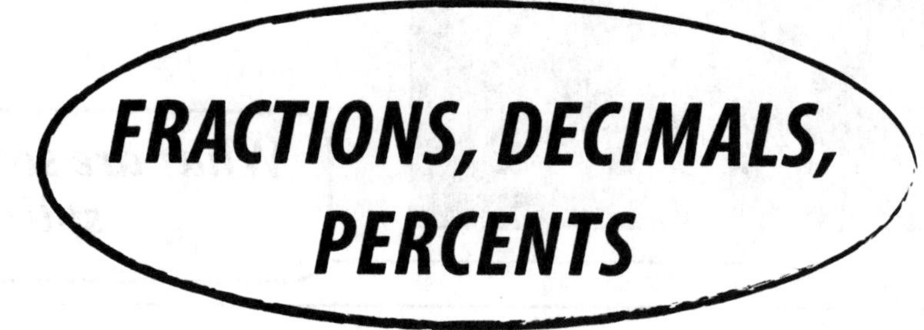

Activity 2: Markdowns

Task 1: Markdown History

Examine the two price tags shown in each of the following problems and complete the stories of the markdowns. Use percents and fractions whenever you can.

1. The original price for the shirt was……

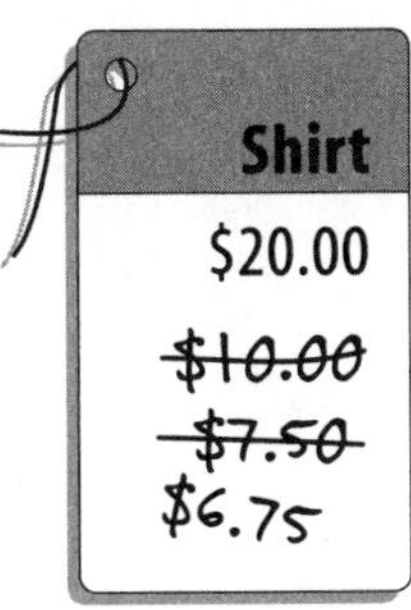

2. First the television cost ……

Task 2: Coupon Sort

Seedlings are normally $2.00 for a six-pack. The nursery sent out eight different coupons to its customers.

1. Sort through the coupons. Put them in order to show the smallest to largest savings. List the card numbers in order here:

2. How did you figure out the order?

3. What questions did you have as you were doing this?

Activity 3: Initial Assessment

Your teacher will show you some problems and ask you to check off how you feel about your ability to solve them. In each case, check off one of the following:

___ Can do ____ Don't know how ____ Not sure

Practice: Coupons in Your Life

Pick a coupon from those that you sorted or a coupon from an actual advertisement that interests you. Attach it to this page.

1. What information do you learn from reading the coupon?

2. Write a paragraph about how you see and use fractions and percents in sales or with coupons.

Split It Up: More Fractions, Decimals, and Percents Unit Goals

What are your goals regarding the study of fractions, decimals, and percents? Review the following goals in this book. Then think about your own goals and record them in the space provided.

- Compare fraction, decimal, and percent amounts to benchmarks such as halves, quarters, and tenths.
- Calculate the percent of an amount by finding 10%, 1%, and their multiples.
- Determine the percent of increase or decrease, given whole numbers.
- Determine a whole amount, given a fraction, decimal, or percent and the part.
- Determine a fraction, decimal, or percent amount for thirds and eighths, given either their percent or fraction forms.
- Choose among fractions, decimals, and percents to solve problems.

My Own Goals

LESSON 1

Numbers in the News

10%, more, or less left to play?

In this lesson, you will focus on **10%** as a **benchmark**, comparing many numbers to determine whether they are more than, less than, or equal to **ten percent**. You will also share ways to find 10% of a total and to determine the total when you know what 10% of it is.

Activity 1: Numbers in the News

1. Write your initials on each card you receive.

2. Compare the amount on each card to the benchmark 10%.

3. Tape each card on the poster where it belongs. If there are cards you are not sure about, put them on the "?" poster.

4. Be prepared to share your reasoning. Make notes in the following table. Think about the *part* and the *whole*.

More Than 10%	Less Than 10%	Equal to 10%	?

5. Complete the following sentence:

   One way to find 10% of an amount is ______________________________

   ____

   ____.

Activity 2: Here Is 10%; What Is the Whole?

1. 10% Now

   a. How many objects do you have? _____ = 10%

   b. How many objects make up the entire amount (the whole)?
   _______ = ______%

   c. Show how you arrived at your answer, using words, pictures, a table, or a diagram.

2. As your classmates report, fill in the first and last columns in the table.

$\frac{1}{10}$, or 10%, of Material for the Design	Rule for Finding the Whole When You Know $\frac{1}{10}$, or 10%	Design Total = $\frac{10}{10}$, or 100%
a.		
b.		
c.		
d.		
e.		
f.		
g.		
h.		
i.		

3. What do you notice as you look across each row of numbers?

4. What is one way to find the whole (100%) when you know the part that is 10%?

5. What words do you associate with 10%? Some examples are "a lot," "almost all," and "not much."

Activity 3: 10%—Yes or No?

1. Do two out of twenty, or $\frac{2}{20}$, equal 10%? Why?

2. Examine the following grids and choose one of them.

   a. Where do you see $\frac{2}{20}$ on your grid?

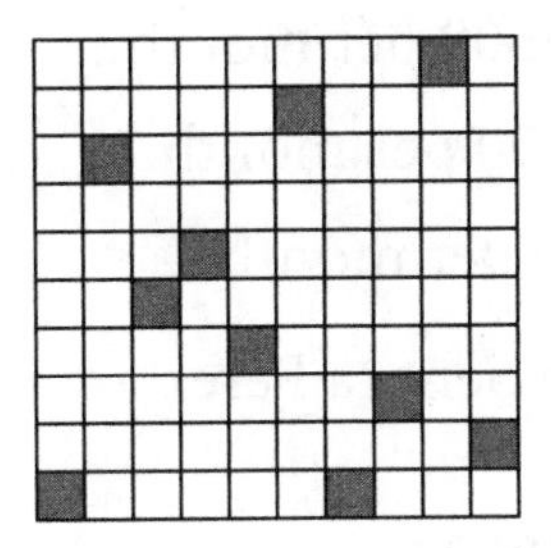
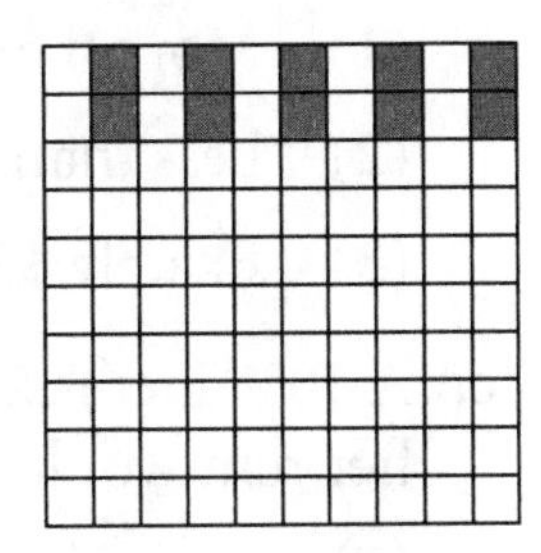
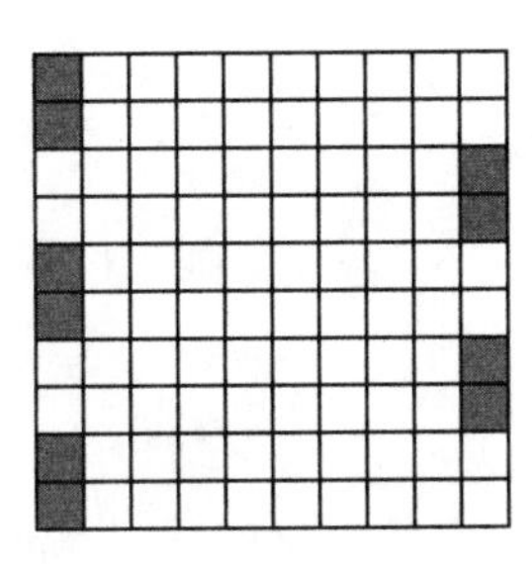

   b. Where do you see 10% on your grid?

3. (Optional) On a calculator, demonstrate step by step the operations that show how $\frac{2}{20} = 10\%$.

Practice: Controlling Costs for Seniors

1. A group of tenants wants to control rent increases. They say a yearly rent increase of 10% is too much. For each situation, circle the answer that correctly describes a 10% rent increase for the tenants.

   a. Aunt Eva's elderly housing costs $80 per month. A 10% rent increase would be

      (1) More than $10.

      (2) Less than $10.

      (3) Equal to $10.

   b. A 10% increase in Monique's rent of $450 each month means she would pay

      (1) More than $500 per month.

      (2) Less than $500 per month.

      (3) Exactly $500 per month.

   c. A 10% increase in Señora Perez' rent of $800 each month means her rent would be

      (1) More than $880.

      (2) Less than $880.

      (3) Equal to $880.

   d. Ten percent of the Davis' rent of $625 per month is

      (1) More than $60.

      (2) Less than $60.

      (3) Equal to $60.

2. A grocery store offers a standard 10% discount for seniors on Wednesdays.

   a. Last Wednesday, Leona's grocery bill before the discount was $72.50. The amount she saved with her senior card was

      (1) More than $10.

      (2) Less than $10.

      (3) Equal to $10.

   b. Leona's total spent for groceries on Saturday was $22.00. If she had waited until Wednesday to shop, she could have saved

      (1) Less than $1.00.

      (2) Between $1.00 and $3.00.

      (3) More than $3.00.

Practice: Money Down

Most banks ask homebuyers to make a down payment of at least 10% of the price of a home when they are taking out a mortgage loan.

1. With a 10% down payment, what price home is each of these families planning on buying?

Family Name	Down Payment of 10%	Home Price
a. Piña	$ 5,000	
b. Kelly	$10,000	
c. Dill	$12,000	

2. Five people in Vernon bought used cars on Valentine's Day. Each person paid 10% in cash and financed the balance. Fill in the missing information on the following chart.

Person	Total Cost	Down Payment of 10%	Balance
a. Ella	$ 650		
b. Irv	$3,300		
c. Bella		$190	
d. Burt		$ 75	
e. Lou			$450

Practice: Drugstore Markups and Markdowns

1. Price adjustments were made at a drugstore. Some items were marked up 10%; others got a 10% markdown. Fill in the chart with the new prices, **rounding** the numbers.

Item	Original Price	Price Change	New Price
a. Gift cards	$1.00	10% markup	
b. Film	$5.00	10% markdown	
c. Soap (travel size)	$0.50	10% markdown	
d. Shampoo	$2.09	10% markup	
e. Mouthwash	$2.99	10% markup	
f. Toothbrush	$3.10	10% markup	
g. Dental floss	$3.99	10% markdown	
h. Bandages	$3.95	10% markup	

2. Cam presented her coupon for 10% off all purchases to the cashier. The cashier totaled her purchases and took off 10%. Cam was not happy and asked for 10% off *each* item. The cashier explained the final total would be the same amount whether she figured it with either method. Who is right? Show how you know.

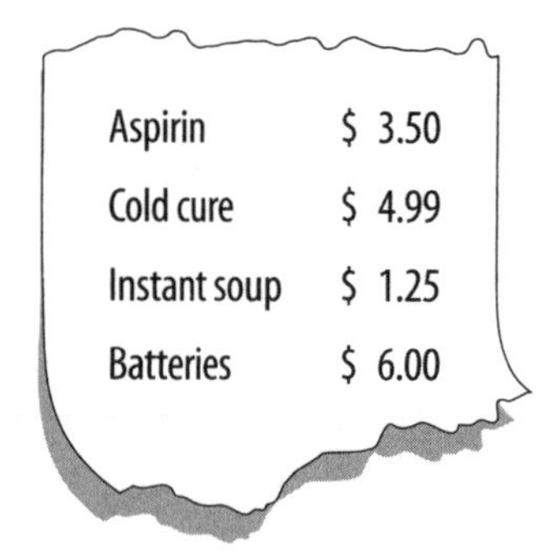

Aspirin	$ 3.50
Cold cure	$ 4.99
Instant soup	$ 1.25
Batteries	$ 6.00

Extension: Cutting Spending

Family A spends $100 at the grocery store. Here is a breakdown of their purchases:

$50.00 on perishables (meat, dairy, produce, etc.)

$10.00 on beverages

$ 5.00 on staples

$ 5.00 on snacks

$10.00 on main meal items

$20.00 on nonfood items (health and beauty, general merchandise)

1. What percent of the total does Family A spend on

a. Perishables? _______

b. Beverages? _______

c. Staples? _______

2. Family B spends $80.00 per week at the grocery store with the same relative amount for each category. How much do they spend on

a. Perishables? _______

b. Beverages? _______

c. Staples? _______

3. Family C spends $120.00 per week with the same percent for each category. How much do they spend on

a. Perishables? _______

b. Beverages? _______

c. Staples? _______

4. Melba is determined to spend less on bills next month so she can save more. Suggest two ways she can cut her spending by 10%.

Melba's Budget

Utility	Monthly Bill	Savings Plan 1	Savings Plan 2
Electricity	$40.00		
Water	$35.00		
Gas	$22.00		
Cable	$43.00		
Phone	$65.00		

Test Practice

1. Which fraction is *not* equal to $\frac{1}{10}$?
   (1) $\frac{100}{1,000}$
   (2) $\frac{20}{200}$
   (3) $\frac{10}{100}$
   (4) $\frac{5}{50}$
   (5) $\frac{2}{200}$

2. By eliminating butter on bread and vegetables, Mary cut 10% of her calorie intake. Mary dropped from 2,200 calories per day to
   (1) 220.
   (2) 980.
   (3) 1,200.
   (4) 1,980.
   (5) 2,000.

3. With a 10% down payment of $7,500, the Cruz family qualified to buy a home of what value?
   (1) $17,500
   (2) $67,500
   (3) $75,000
   (4) $750,000
   (5) $7.5 million

4. Average rent for a small apartment dropped from $900 to $850. The percent of decrease is
   (1) Less than 10%.
   (2) Exactly 10%.
   (3) Between 10% and 25%.
   (4) Exactly 25%.
   (5) More than 25%.

5. Phyllis, Ivy, and some friends at work won an office lottery. They decided that they would *not* share the money equally. Phyllis won $\frac{1}{9}$ of the money, and Ivy won $\frac{1}{11}$. Which statement is true?
   (1) Phyllis got less than 10%.
   (2) Ivy got more than 10%.
   (3) Phyllis got less money than Ivy.
   (4) Phyllis got more money than Ivy.
   (5) We need to know how much money they each won in order to tell who got more.

6. Molly withdrew cash from an ATM. She put 10%, or $15, in her pocket and hid the rest. How much money did she withdraw in total?

LESSON 2

What Is Your Plan?

How could you use percents to plan the space?

In this lesson, you will explore common percents in a variety of problems, including one in which you will create a plan for the use of display space in a market stall. Common percents include **multiples of 10%** such as 20%, 30%, 40%, 50%, 60%, 70%, 80%, and 90%, as well as the benchmarks **25%** and **75%**.

Having a clear picture and understanding of common benchmark percents is critical to understanding data and statistics. It is also useful in your everyday life when you are figuring the cost of a sale item, enlarging or reducing a copy, or allocating space for specific purposes.

Activity 1: Display Plans

You run a farmer's market stall on fall weekends. Your stall has 50 square feet of display space that measures 10 feet by 5 feet.

You use

- 40% of the space to display apples;
- 30% of the space to display pears;
- 20% of the space to display grapes; and
- 10% of the space to display jams.

1. Use the percents to make a plan.

On the grid paper your teacher will give you, show how you could arrange your stall. Then answer the questions.

When you are done with your Fresh Fruit display plan and questions, put your initials on the back and post your plan.

Review the plans others have posted, and look at the ways that they found 20% and 30%. If you have a question about or disagree with a particular plan, tag the plan with a question mark (?) on a Post-it® Note.

2. Key (an explanation of what the different colors or letters in the squares represent):

3. How many square feet are needed to show the 20% of space used for displaying grapes? How do you know? Explain how you know with words, pictures, and numbers.

4. How many square feet are needed to show the 30% of space used for displaying pears? How do you know?

Activity 2: What Is *Your* Plan?

Choose one plan to design. Circle your choice.

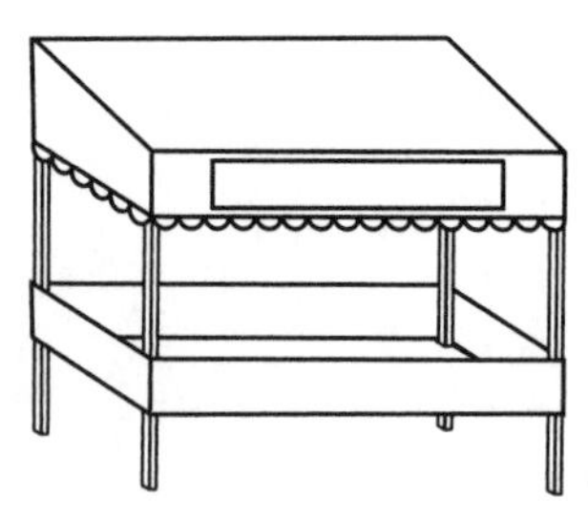

Snack pushcart

Souvenir pushcart

Children's game pushcart

Newspaper/magazine pushcart

Your display area measures **10 feet by 5 feet (50 square feet)**.

You sell three or four types of items. What percent of your space will be used for each type of item?

_____% will be used to display Item A: ____________________________

_____% will be used to display Item B: ____________________________

_____% will be used to display Item C: ____________________________

_____% will be used to display Item D: ____________________________

Use another grid to show your pushcart plan, and answer the following questions.

1. My plan used _____ squares.
2. The key for my plan:

   Item A:

   Item B:

   Item C:

   Item D:

3. What was the total percent of space you used to display items? Show how you know.

4. Write a fraction (part/whole) that describes the squares that equal the percent for each item in your plan. How can you prove the percents and fractions are equal?

 a. Item A: ______% ______fraction of squares

 b. Item B: ______% ______ fraction of squares

 c. Item C: ______% ______ fraction of squares

 d. Item D: ______% ______ fraction of squares

Activity 3: 20% Dilemma

If 10% is one-tenth ($\frac{1}{10}$), is 20% one-twentieth ($\frac{1}{20}$)?

1. Use the following grid to prove your answer.

2. List some other fractions for 5%.

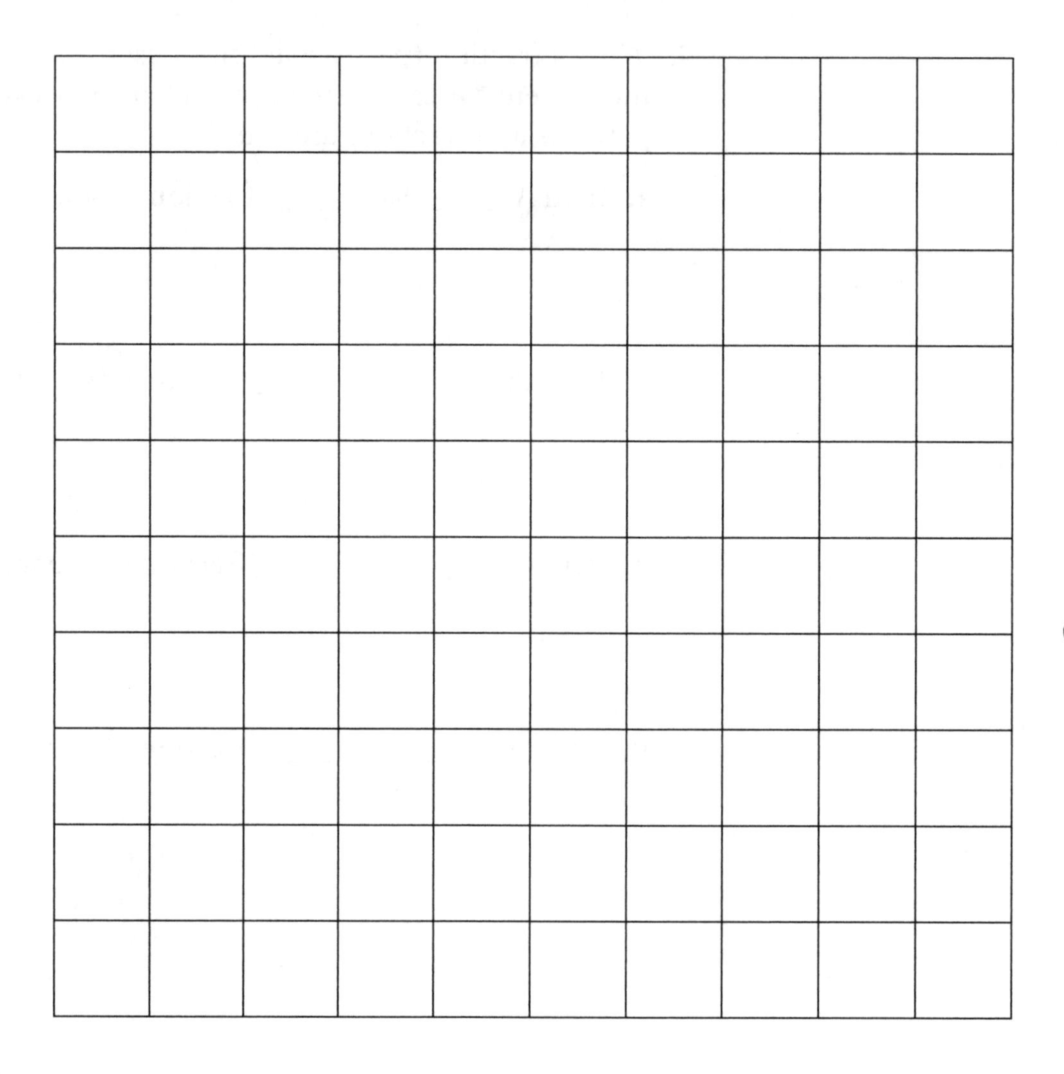

Practice: More Plans

> If you use *all* your space, what percent will you use? How do you know if your percents total the right amount? How many squares are in the grid? What percent does each square represent?

Design your own floor plan on the following grid:

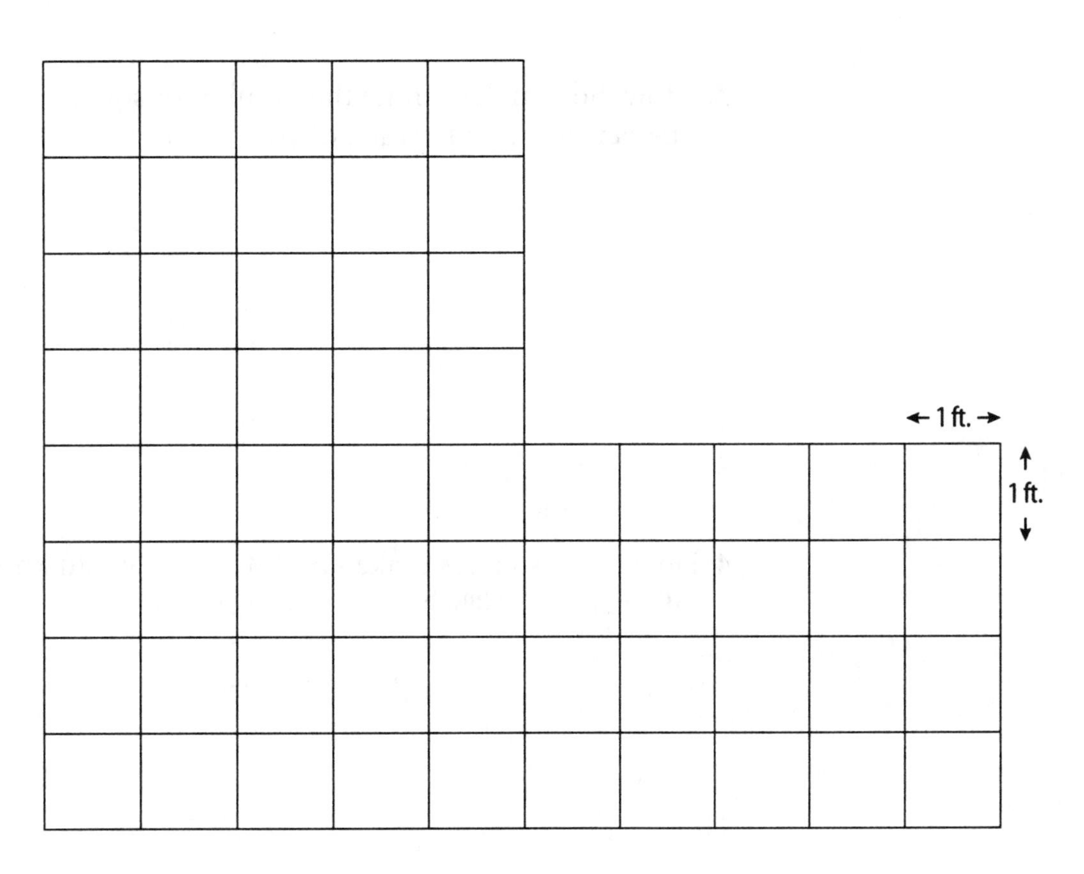

My floor plan shows a ______________________________.

1. List the percent of space you will use for each item. You must show *at least three* different items on your floor plan.

   _____% is ____.

   _____% is ____.

   _____% is ____.

   _____% is ____.

2. Key (what the different squares represent):

3. How did you determine the number of squares needed to represent the percent used for each of your items?

4. How many squares make up 35%? How would knowing the number of squares in 10% help you answer this?

Practice: Visual Percents

1. Select the percent that best matches each circle.

Think about benchmark percents: 25%, 50%, 75%, and 100%.

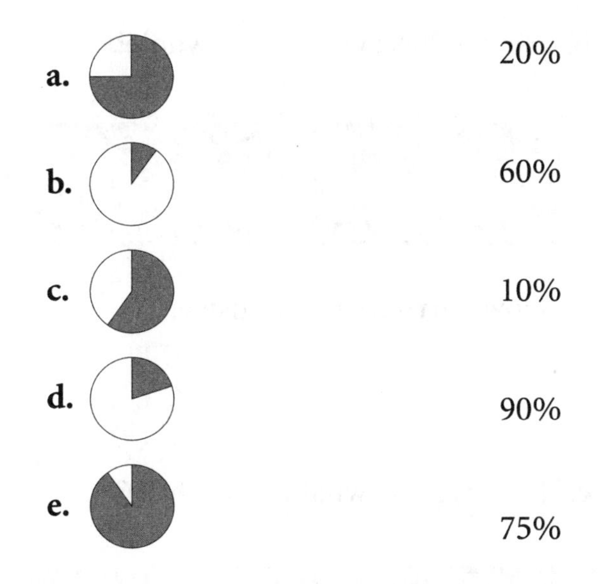

2. On the following graph, mark the vertical axis in increments of 10%, starting at zero and going up to 100%. Remember the spaces between the percent increments must be even!

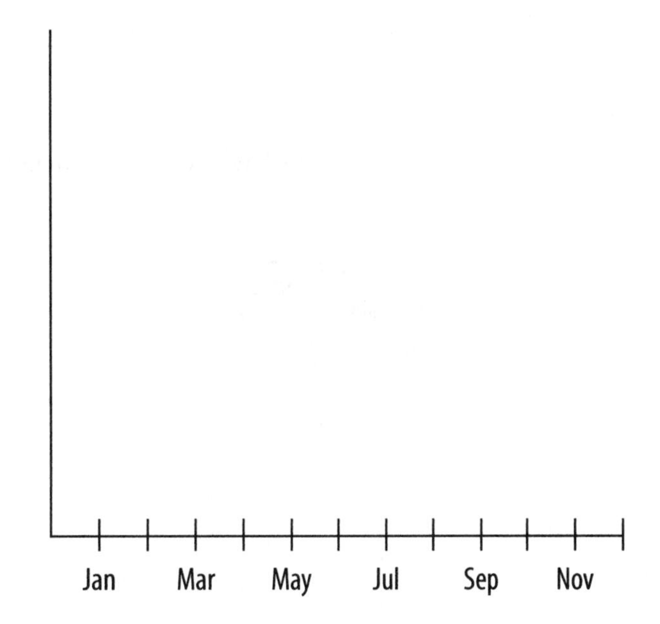

3. For each of the following problems, show the whole. For Problems b and c, explain how you arrived at your answer.

   a. Here is 10%; what is the whole?

   b. Here is 20%; what is the whole?

   How I arrived at my answer:

   c. Here is 30%; what is the whole?

   How I arrived at my answer:

   d. Here is 70%; what is the whole?

4. Mark each of the line segments with the following percents and the corresponding amounts they represent:

10% 20% 25% 50% 70% 90%

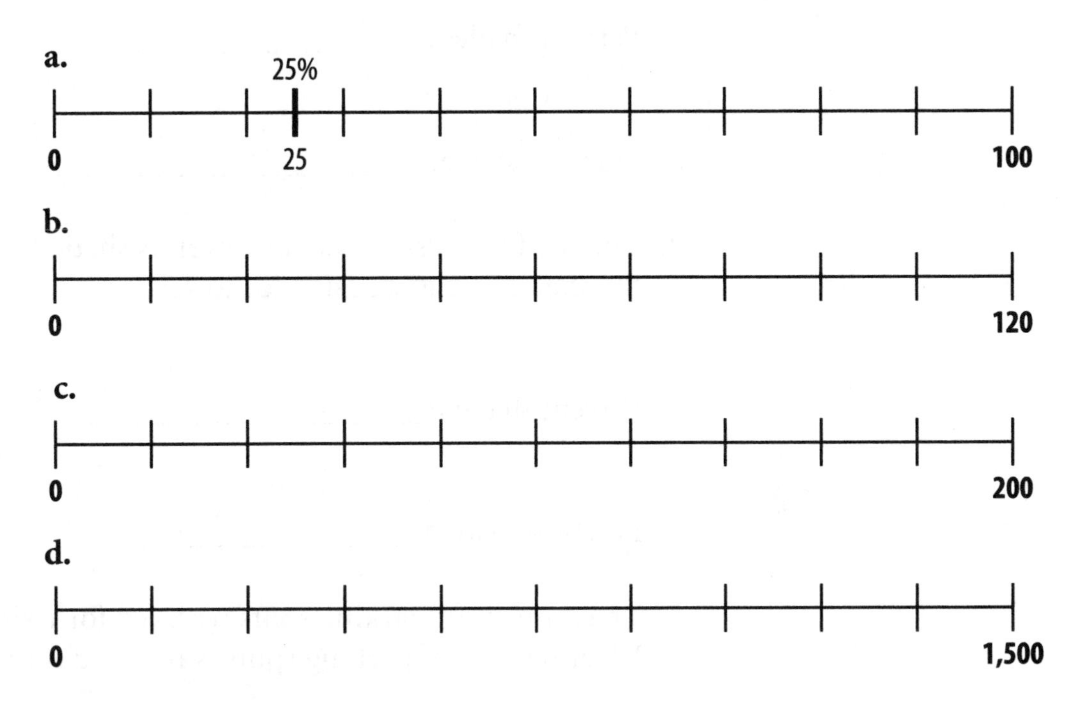

5. Find the percent and fraction shaded for each of the following problems.

   a. The area in this suburb slated for development is shaded.

   Percent shaded: ____________________

   Fraction shaded: ____________________

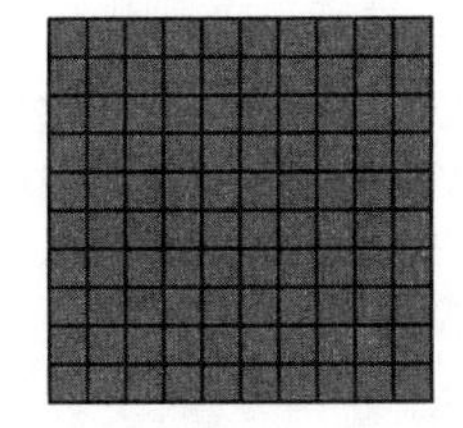

   b. The number of handicapped accessible apartments in this building is shaded. What percent are handicapped accessible?

   Percent shaded: ____________________

   Fraction shaded: ____________________

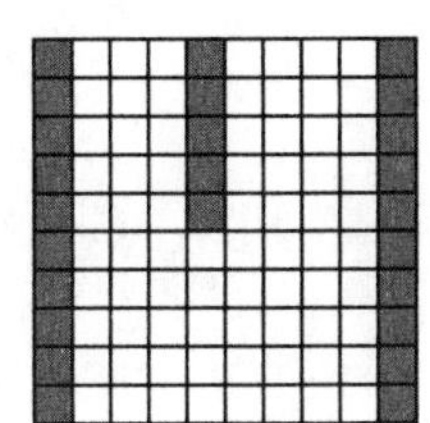

c. The swimming lanes open to the public are shaded. What percent are open to the public?

Percent shaded: ____________________

Fraction shaded: ____________________

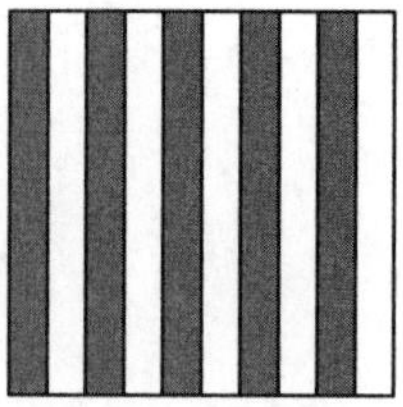

d. The shelf space set aside for towels is shaded. What percent of the shelf space is set aside for towels?

Percent shaded: ____________________

Fraction shaded: ____________________

e. The number of parking spots reserved for visitors is shaded. What percent of parking spots is reserved for visitors?

Percent shaded: ____________________

Fraction shaded: ____________________

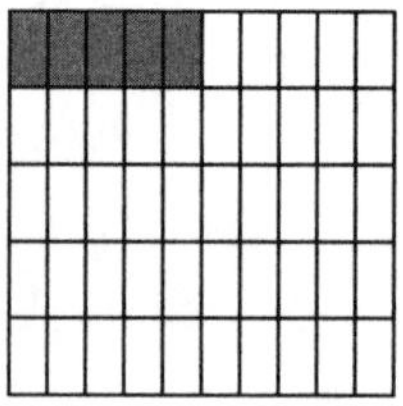

Practice: Comparing Percents

Following are three symbols used to compare numbers:

> (is greater than)   < (is less than)   = (equals)

Remember that *the pointed end of the symbol always points to the smaller amount.*

Use these symbols between the amounts described and the percents listed to make true statements.

Example:

Amount	Comparing Symbol	Percent
20 beets picked; 8 eaten	>	30% eaten

Reason: 10% of 20 beets is 2 beets, 20% is 4 beets; 30% is 6 beets; 40% is 8 beets — 40% is greater than 30%.

1. 20 miles to travel;
   14 miles covered — 80% covered
2. 30 test questions;
   20 questions correct — 70% correct
3. 50 pennies in a roll;
   25 pennies rolled — 40% rolled
4. 150 calories per glass of milk;
   30 calories consumed — 20% consumed
5. 300 hours of work;
   275 hours paid for — 90% paid for
6. $1,200 bonus;
   $800 spent — 70% spent
7. $2,500 savings goal;
   $800 saved — 30% saved

8. Explain your choice of symbols for Problems 4 and 6. (See the *example* for ideas.)

a. Problem 4:

b. Problem 6:

Test Practice

1. Eduardo bought a shirt and socks, originally marked $14.99 and $4.99. If both items were 20% off, how much did he pay for both of them? (Round to the nearest dollar.)

   (1) $2

   (2) $4

   (3) $16

   (4) $18

   (5) $20

2. One-fifth of $1,000 is

   (1) The same as 5% of $1,000.

   (2) The same as 10% of $1,000.

   (3) The same as 15% of $1,000.

   (4) The same as 20% of $1,000.

   (5) The same as 50% of $1,000.

3. Clara works 40 hours a week. She spends 16 of the 40 hours on the phone. What percent of her work time is phone time?

   (1) 10%

   (2) 25%

   (3) 30%

   (4) 40%

   (5) 50%

4. Tito is a forest ranger. He is asked to map an area of the state forest to keep track of wildlife. He marks off an area equivalent to 20% of the whole forest and finds that six deer roam in that area. If the deer population is evenly distributed around the park, what is the total number of deer Tito will predict live in the park?

   (1) 6

   (2) 12

   (3) 20

   (4) 24

   (5) 30

5. The town of Hooperville has 1,200 households. Seventy percent of those households include children. How many Hooperville households have no children?

   (1) 120

   (2) 360

   (3) 840

   (4) 1,200

   (5) 8,400

Item 6 is based on United States Census Bureau information compiled for the years 2000, 2001, and 2002, and then averaged.

6. According to the U.S. Census Bureau, on average just under 20% of people in Arkansas (18.8%) and Mississippi (18.9%) live in poverty. In one town of 5,000 people, 20% of the population lives in poverty. How many people are poor in that town?

LESSON 3

One Percent of What?

Is 10% always larger than 1%?

Sometimes you need to know what **1%** of an amount equals. For example, you might hear that "only 1% of the raffle tickets are winners," "**one percent** of the marathon runners finished the race in less than two-and-a-half hours," or "only 1% of the votes went to your candidate in the primaries." In each case, 1% will be a different amount, depending on the total—1% of what?

In this lesson, you will think about ways to find and name 1%.

Activity 1: Go Figure!

One day Jamal declared, "Ten percent is always larger than 1%." Juan said, "I don't think so." Who is right?

Compare the discounts:

Use number lines, grids or diagrams, objects, or a table to prove your solution. Using more than one method works well as a way to double check your answer!

Demonstrate your solution below:

Activity 2: Patterns with 10% and 1%

Three-Digit Numbers

Percent	Number ____	Number ____	Number ____	Number ____	Number ____
100%					
10%					
1%					

Four-Digit Numbers

Percent	Number ____	Number ____	Number ____	Number ____	Number ____
100%					
10%					
1%					

Two-Digit Numbers

Percent	Number ____	Number ____	Number ____	Number ____	Number ____
100%					
10%					
1%					

1. Explain the pattern you notice when you look down the columns of each table.

2. What causes the numbers to change as they do?

Activity 3: Mental Math Comparisons

Compare each pair of amounts. Do the math in your head. Use the benchmarks 10% and 1%.Then use the symbols < (less than), = (equals), or > (greater than) to show the comparisons.

Make notes about how you reasoned.

1. 20% of $500 ___ 2% of $3,000

20% of $500 = _______

2% of $3,000 = _______

2. 40% of $800 ___ 4% of $9,000

40% of $800 = _______

4% of $9,000 = _______

3. 30% of $1,500 ___ 3% of $10,000

30% of $1,500 = _______

3% of $10,000 = _______

4. 60% of $200 ___ 6% of $2,000

60% of $200 = _______

6% of $2,000 = _______

5. 50% of $900 ___ 5% of $10,000

50% of $900 = _______

5% of $10,000 = _______

6. 70% of $1,200 ___ 7% of $12,000

70% of $1,200 = _______

7% of $12,000 = _______

Practice: Which Is Greater?

Use what you know about finding percents to determine whether the amount in the left or the right column is greater. Check the *greater* amount. One is done for you as an example.

Left Column Amounts	Right Column Amounts
1% of 5,000 troops = 50 troops	10% of 900 troops = 90 troops ✓
1. 10% of 100 steps =	1% of 1,200 steps =
2. 1% of 3,000 voters =	10% of 400 voters =
3. 1% of 2,700 mill workers =	10% of 250 mill workers =
4. 10% of 750 pages =	1% of 1,150 pages =
5. 10% of 10 games =	1% of 150 games =
6. 10% of 35 TV shows =	1% of 450 TV shows =
7. 1% of 10,000 viewers =	10% of 1,500 viewers =
8. 1% of 3,850 strikers =	10% of 390 strikers =

9. How did you find 10% of the numbers in the chart above?

10. How did you find 1% of the numbers in the chart above?

11. What shortcuts, if any, do you use when comparing 1% and 10% of different numbers?

Practice: Fundraisers

The local youth club holds six fundraisers each year. Club members decided to set aside 1% of the proceeds from each event for emergencies.

1. How much money should the youth club treasurer report was added to the emergency account for the year?

Youth Club Fundraising Report

Fundraising Event	Gross Receipts	1% Emergency Fund	10% Cost of Fundraiser	____% Remaining Receipts
Bake Sale	$ 800			
Car Wash	$ 600			
Raffle	$3,150			
Bumper Sticker Sales	$ 250			
Craft Fair	$1,200			
Walk-a-thon	$1,800			
Total				

2. Complete the table.
3. What *percent* must be recorded at the top of the last column?

4. How can you use the table to check the math?

Practice: Growing Cities and Towns

Cities and towns grow and shrink in population. How would a 1% increase in population change these towns' sizes?

1. Fill in the following table.

City/Town	1990 Population (rounded to nearest 100)	1% of Population	New Population If Increased by 1%
Amador	30,000		
Del Norte	24,000		
Glenn	25,000		
Mono	10,000		
San Diego	2,498,000		
San Francisco	724,000		
Sierra	3,300		

Data from Counting California: http://countingcalifornia.cdlib.org

2. List some changes that occur with a population increase.

Extension: Working Backward

Sometimes you know how much a part of a whole is, but not how much the whole is. In the following examples, you are given an amount equal to 1% of a whole. Use the clues to figure out 100%, the whole.

How does 10% of an amount compare to 1% of the same amount?

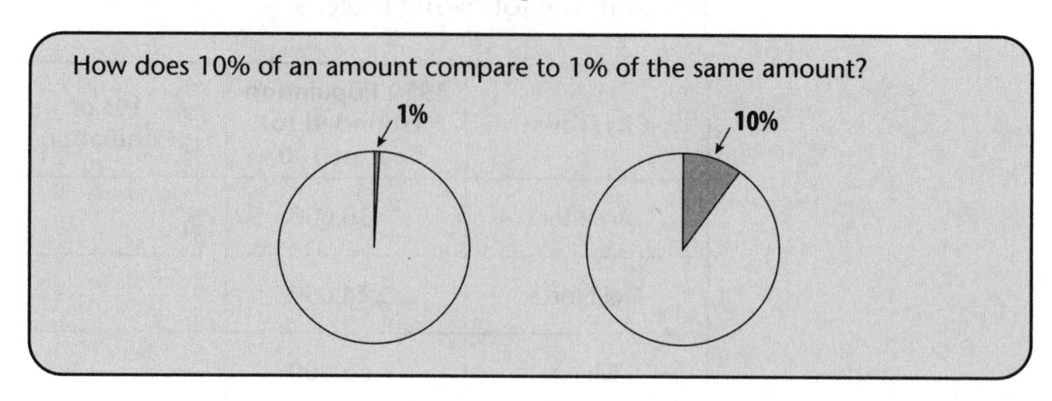

1. About 1% of my daily calories come from eating chocolate. I eat about 20 calories of chocolate each day. How many total calories do I eat each day?

1%	10%	100%
20 calories		

2. One percent of all voters are newly registered in Hobbesville. There are 57 newly registered voters in town. What is the total number of voters in Hobbesville?

1%	10%	100%

3. About 1% of the cookies produced at Sweet Spirit Bake Shop are unusable. If Sweet Spirit Bake Shop produced 18 unusable cookies yesterday, what was the total cookie production for the day?

4. Lila devotes 1% of her monthly work hours to answering phones. Last month she logged 1.5 hours of phone time. How many total hours did she work last month?

Test Practice

1. Which of the following numbers does *not* represent 1%?

   (1) $\frac{2}{200}$

   (2) 0.1

   (3) 0.01

   (4) .01

   (5) $\frac{1}{100}$

2. One-hundredth of $1,000 is the same as

   (1) 1% of $10,000.

   (2) 10% of $1,000.

   (3) 10% of $10,000.

   (4) 20% of $2,000.

   (5) 50% of $20.

3. Waiters at Davenport's Restaurant each set aside 10% of their tips to give to the busboy. The waiter Akeem collected $150 in tips one night. How much tip money did he take home?

   (1) $1.50

   (2) $15.00

   (3) $135.00

   (4) $148.50

   (5) $151.50

4. Tony earns 1% interest on his checking account. Last month he earned $3 in interest. How much did he have in his account?

   (1) $3

   (2) $30

   (3) $300

   (4) $3,000

   (5) $30,000

5. Mimi signs up to contribute 1% of her earnings toward a retirement plan. To figure out how much she needs to contribute each week, Mimi might do any of the following *except*

   (1) Find 10% of her weekly earnings and then find 10% of that number.

   (2) Divide her weekly earnings by 10 and then divide by 10 again.

   (3) Divide her weekly earnings by 100.

   (4) Multiply her weekly earnings by 0.1.

   (5) Multiply her weekly earnings by 0.01.

6. Twenty-five percent of $600 is the same as 10% of ________.

LESSON 4

Taxes, Taxes, Taxes

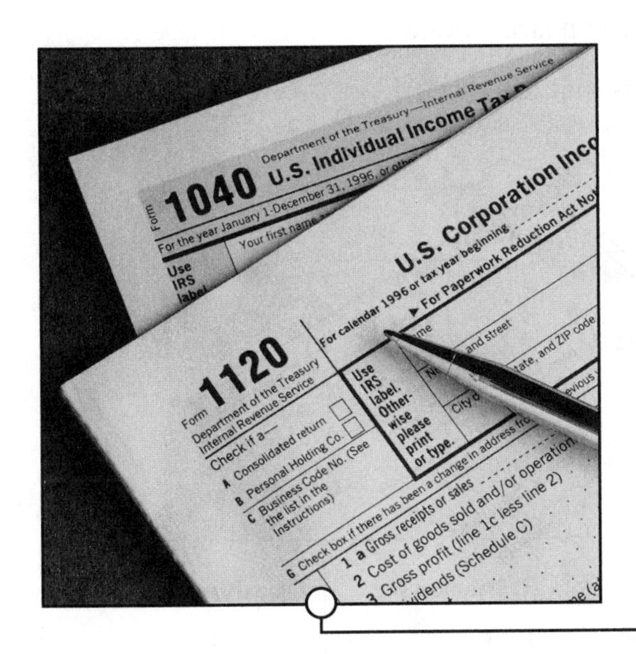

How is the tax determined?

An old saying goes, "Two things are certain in life: death and taxes." Taxes cost individuals and companies money, but paying taxes gives the government revenue to provide services. Understanding how to calculate taxes can help you keep track of where your money goes.

In this lesson, you will use what you know about finding **multiples of 1%** and 10% to determine the sales tax for various items and to compare wages.

Activity 1: Different States, Different Charges

In 2004, only five states out of 50 did not charge sales tax. The other 45 states charged different rates, from a low of 2.9% in Colorado to a high of 7.25% in California. Most states charged in the range of 4%–7% for sales tax.

- Fill in the following chart.
- Round the item prices to the nearest dollar.
- Then answer the questions that follow.

Sales Tax Table

Sales Tax Rate*	Car Stereo $_____	Sneakers $_____	Washing Machine $_____
	Tax Paid on Items		
4% AL, GA, HI, LA, SD, WY			
5% IA, ME, MD, MA, NM, ND, SC, WI			
6% CT, FL, ID, IN, KY, MI, NJ, OH, PA, VT, WV			
7% MS, RI, TN			

*States charging this percent sales tax in 2004

1. What are *two* ways you can figure the amount charged when there is a 5% sales tax on an item?

2. What is the final cost of the car stereo, including the sales tax, in Florida (FL)? How do you know?

3. How much more does a stereo cost (including the sales tax) in Michigan (MI) than in Alabama (AL)? How do you know?

4. What is the difference in price for a washing machine (including the sales tax) between states listed with the lowest sale tax and those with the highest sales tax?

Activity 2: Take-Home Pay

1. When Mara moved to Oregon, she earned four times as much as her brother in Louisiana earned. She figured that she would take home four times as much too. Do you agree? Why or why not?

Gross Weekly Pay	Federal* Tax	State Tax	Social Security	Medicare	Net Pay
Mara's: $1,600	(28%)	(9%—OR)	6%	1%	
Her brother's: $400	(15%)	(2%—LA)	6%	1%	

*All tax figures are based on 2003 tax rates and are rounded to whole numbers.

Based on the tax figures (from 2004 tax tables), Mara compared their take-home, or net, pay. What did she discover?

2. Mara's take-home pay:

3. Her brother's take-home pay:

4. When Mara compared them, she found that …

Practice: Personal Payroll Deductions

A paycheck stub lists the **gross pay** (the total amount) and the take-home, or **net**, pay after taxes and deductions for benefits.

Use the following information to determine what the take-home amount would be after all deductions.

Show your work.

1. Fill in your own weekly gross pay or make up an amount.

   Gross weekly pay (hours worked x hourly wage): __________

2. The percent withheld for Federal income tax: _______
   (Choose the category that applies to you.)

3. The amount withheld for Federal income tax per week from your gross pay: _______

Weekly Withholding for Federal Income Tax

Tax Rate*	Single	Married Filing Jointly	Married Filing Separately	Head of Household
10%	Up to $134.62	Up to $269.24	Up to $134.62	Up to $192.31
15%	$134.63–$546.15	$269.25–$1,092.31	$134.63–$546.15	$192.32–$731.73
25%	$546.16–$1,323.08	$1,092.32–$2,204.81	$546.16–$1,679.81	$731.74–$1,889.42

*All tax rates are based on 2003 figures.

Check **http://www.taxadmin.org** or call the reference librarian at your local library for the state income tax rates in your state.

4. The percent withheld for state income tax in your state: _______

5. The amount withheld for state income tax per week from your gross pay: _______

Other Deductions

Social Security tax rate for employees: 6% (approximate)

Medicare tax rate for employees: 1% (approximate)

6. The percent withheld for other deductions per week from your gross pay: ______
7. The amount withheld for other deductions: ______
8. Net weeky pay: ______

Practice: Which Is a Better Deal?

Barker's Bargain Basement and Elsa's Emporium are department stores at West Side Mall. They sell a lot of the same things, but they price them differently.

Both stores are having big sales. For each item, circle the better deal. Show with pictures, words, or numbers how you found your answers for Problems 2 and 5.

For Problem 6, choose your own "___ %-off" sale for each stapler and show what the final cost would be.

1. Jacket

At Barkers:

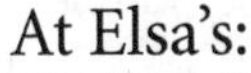

At Elsa's:

2. Set of Kitchen Knives

At Barkers:

At Elsa's:

3. Portable CD Player

At Barkers:

At Elsa's:

4. Running Shoes

At Barkers:

At Elsa's:

5. Dollhouse

At Barkers:

At Elsa's:

6. Stapler

At Barkers:

At Elsa's:

Practice: Increase, Decrease

Match the descriptions on the left with the appropriate amounts listed on the right.

Descriptions	Amounts
1. Population: 3,000; increased by 35%	9,520
2. Manufacturing jobs: 1,200; decreased by 8%	731.5
3. Fundraising target: $8,500; increased by 12%	1,160
4. Registered voters: 800; increased by 45%	8,610
5. Technology costs: $10,500; decreased by 18%	4,050
6. Carpet coverage: 950 sq. ft.; decreased by 23%	1,104

7. Cassie figured the total for Problem 5 to be $8,190, *not* $8,610. Check her math. What method did she use to solve the problem?

- Explain what each number is.
- Circle and explain the error(s).

Cassie started with $10,500.

Step 1	Step 2	Step 3
$1,050	$2,100	$10,500
$1,050	+ 210	– 2,310
$ 105	$2,310	$ 8,190
$ 105		

Practice: Visuals of Percents

Each of the following diagrams represents a percent of a whole diagram. Draw or describe in words what 100% of each diagram would look like.

1. Twenty-seven percent of the circle is shaded.

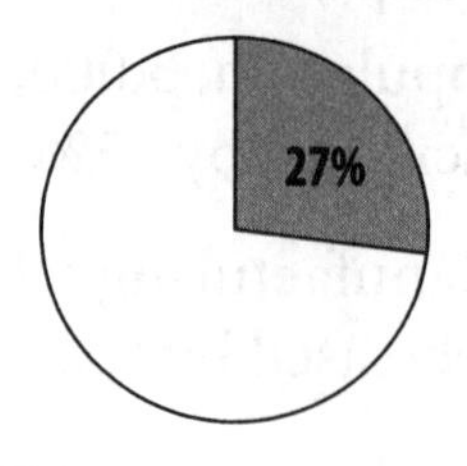

a. Shade the rest.

b. How do you know 100% of the circle is now shaded?

c. What percent did you shade? ________

2. This is 15% of a line of dots. How many dots represent 100% of the line? How do you know?

3. **a.** What percent of the line below is drawn? ________

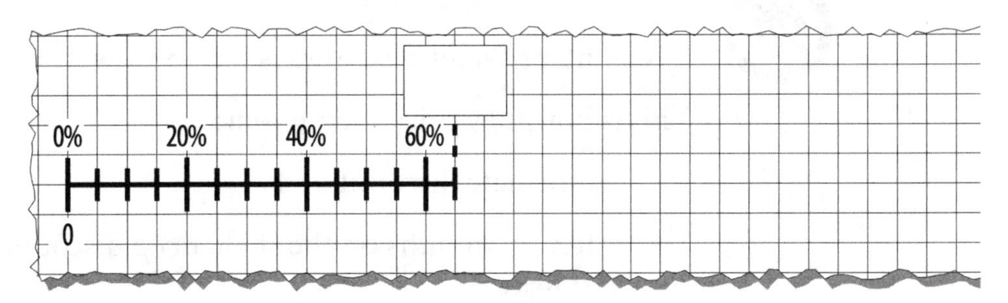

b. Draw the line out to 100% of its length.

c. What percent of the line did you draw? ________

d. How do you know your line equals 100% of the length?

4. This grid represents 25% of a whole.

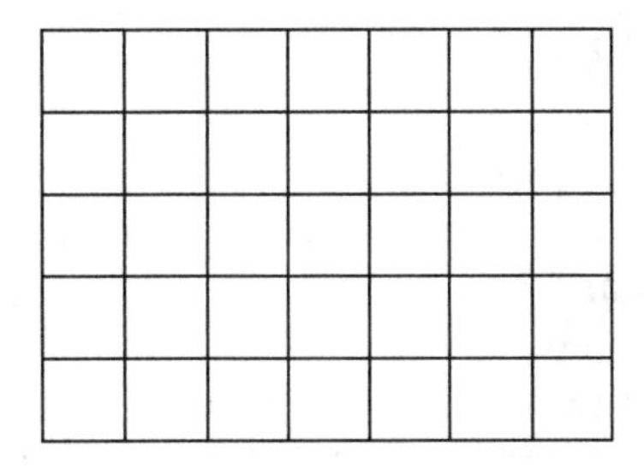

a. Show a shape that would represent 100% of a whole.

b. How do you know your shape equals 100% of a whole?

Extension: Markdowns

Discounter's Warehouse always has the lowest prices in town.

This is how they price their goods:

- Immediate discount of 4%
- After six months on the shelf, deep discount of 35% off the *last* price
- After 12 months on the shelf, a deeper discount of 12% off the *last* price

1. Fill in the markdown history of the items in the following chart. You may want to use a calculator. Round your answers to the nearest cent.

Discounter's Warehouse Markdown Schedule

Initial Price	Immediate Discount		Deep Discount (after 6 months)		Deeper Discount (after 12 months)	
	4% Discount	Price	35% Discount	Price	12% Discount	Price
Winter Jacket $125.00						
Standing Lamp $44.00						

2. Would it be a better deal for the consumer if Discounter's Warehouse gave a 51% markdown off the original price instead of using the markdown schedule? Explain.

Test Practice

1. The American Heart Association recommends that saturated fat intake be limited to 300g each day. Four ounces of dark-meat chicken with skin contains 15 grams of saturated fat, which is

    (1) 1% of the recommended daily saturated fat intake.

    (2) 5% of the recommended daily saturated fat intake.

    (3) 10% of the recommended daily saturated fat intake.

    (4) 15% of the recommended daily saturated fat intake.

    (5) 20% of the recommended daily saturated fat intake.

2. An analysis of supermarket sales showed that chocolate ice cream accounted for 19% of all ice cream sales. To determine the amount of chocolate ice cream a supermarket should order, you could do any of these calculations *except*

    (1) Find 10%, 5%, and 4% of your ice cream order.

    (2) Find $\frac{19}{100}$ of your ice cream order.

    (3) Find $\frac{1}{5}$ of your order and subtract 1%.

    (4) Find one-quarter of your order and add 1%.

    (5) Find 20% of your order and subtract 1%.

3. A supermarket usually orders 300 gallons of ice cream each month. Chocolate ice cream accounts for 19% of the order. How many gallons of chocolate ice cream does the supermarket order? [Find 10%, 5%, and 4% of your ice cream order, or subtract 1% from 20%].

    (1) 63 gallons

    (2) 60 gallons

    (3) 57 gallons

    (4) 27 gallons

    (5) 18 gallons

4. An analysis of supermarket sales showed that vanilla ice cream accounted for 45% of all ice cream sales. According to the graph, which store's ice cream sales were at least 45% vanilla ice cream?

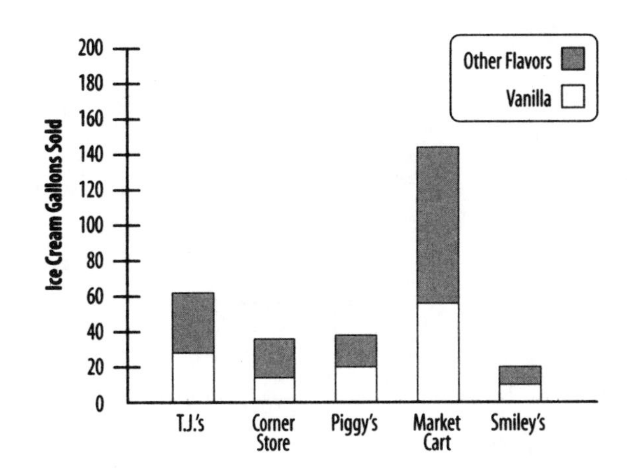

    (1) Market Cart and T.J.'s

    (2) Corner Store and Smiley's

    (3) Corner Store and Piggy's

    (4) Corner Store and Market Cart

    (5) Piggy's, Smiley's, and T.J.'s

5. The percent of nonmanagerial workers who get paid for sick days dropped from 67% in 1997 to 60% in 2002. The Bright Company had 1,000 nonmanagerial workers in both 1997 and 2002. How many *fewer* nonmanagers were likely paid for sick days in 2002?

   (1) 60

   (2) 70

   (3) 330

   (4) 600

   (5) 670

6. In Massachusetts in 2003, 48% of GED certificates were awarded to teens. If 9,700 Massachusetts residents received a GED certificate, how many were teens?

LESSON 5

Fold and Figure

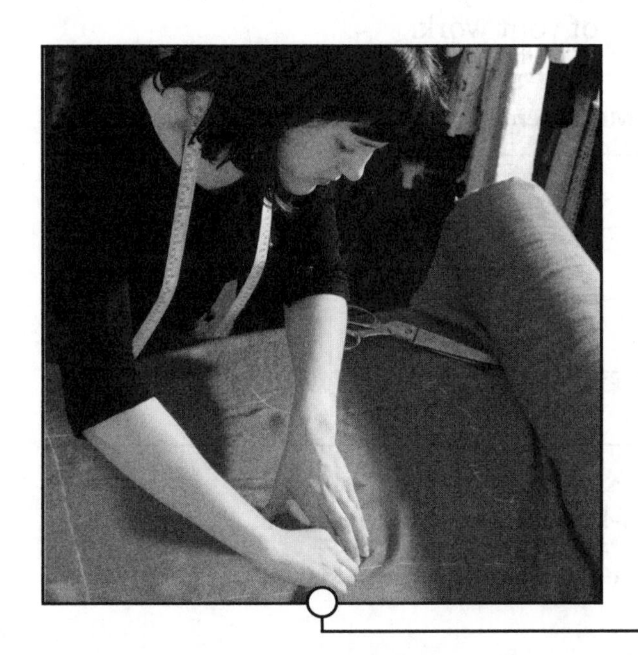

How much is one-eighth of a yard?

You have worked with benchmark percents, such as 10% and 1%, and the fractions $\frac{1}{2}$ and $\frac{1}{4}$. Now you will focus on $\mathbf{\frac{1}{8}}$.

In this lesson, you will measure various items and consider high-interest loans to explore **one-eighth** as a fraction and as a percent. You will also shade grids to relate the fraction and percent to a decimal.

Activity 1: Fractions of Yards

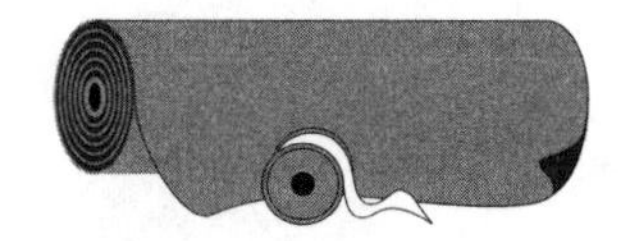

Fabric samples and ribbons are sometimes sold in $\frac{1}{8}$-yard lengths. How long is $\frac{1}{8}$ of a yard? How long is $\frac{2}{8}$ or $\frac{3}{8}$?

1. Measure the ribbons posted around the room, and complete the following table. Keep track of your work.

Item	Measurement in Inches	Measurement in Yards
A		
B		
C		
D		

2. Mark off eighths on a yard-long strip of masking tape: $\frac{1}{8}$, $\frac{2}{8}$, $\frac{3}{8}$, $\frac{4}{8}$, $\frac{5}{8}$, $\frac{6}{8}$, $\frac{7}{8}$, and $\frac{8}{8}$. Draw a picture to represent your marked tape.

3. Determine the length in inches at each mark:

   a. $\frac{1}{8}$ yd. = _____ in.

   b. $\frac{2}{8}$ yd. = _____ in.

   c. $\frac{3}{8}$ yd. = _____ in.

   d. $\frac{4}{8}$ yd. = _____ in.

   e. $\frac{5}{8}$ yd. = _____ in.

   f. $\frac{6}{8}$ yd. = _____ in.

   g. $\frac{7}{8}$ yd. = _____ in.

   h. $\frac{8}{8}$ yd. = _____ in.

4. Familiar Fractions:

   a. You bought $\frac{1}{2}$ yard of silk; how long in inches is your piece of fabric?

   b. You bought $\frac{1}{4}$ yard of cotton; how long in inches is your piece of fabric?

   c. You bought $\frac{3}{4}$ yard of wool; how long in inches is your piece of fabric?

   d. You bought one yard of nylon; how long in inches is your piece of fabric?

5. a. If you cut your $\frac{1}{8}$ yard of ribbon in half, what fraction of a yard would you have? How do you know?

   b. How many inches long would the ribbon piece be? How do you know?

   c. What fraction of a yard would you have if you cut this new piece of ribbon in half? How do you know?

6. The following line represents one piece of yarn. Show how you would cut it in half, then cut that half in half again, then cut it in half again, then in half again, and then in half again. Label each cut mark with a fraction. *Note*: You will have five lines and five fractions on your piece of yarn.

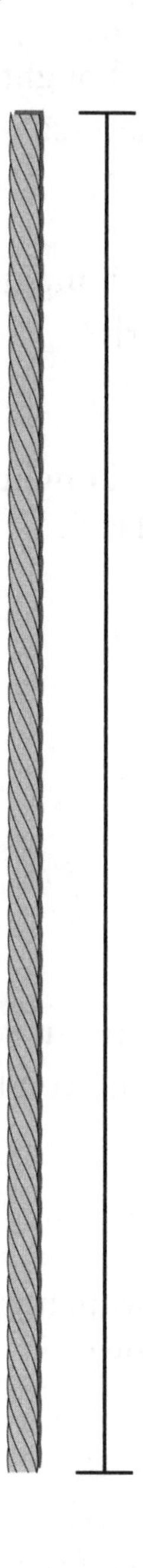

Activity 2: The Loan Shark

Manny needs to borrow $2,400 for a car. His credit is bad, so he visits Louis, the loan shark. Louis says, "It's your lucky day. I'm only going to charge you $12\frac{1}{2}$% interest. That's a bargain! Take it before I change my mind."

1. Manny has to decide quickly. He knows how to find 10% and 1% of a number easily, so he figures out what $12\frac{1}{2}$% (12.5%) interest on his $2,400 loan would be. He makes a few notes on paper. What numbers might he write down as he tries to figure out the interest amount?

2. Manny starts to doubt his math, so he checks the numbers a second way. How might he have checked his answer?

Manny's brother, Sammy, has a terrible credit rating. When he visits Louis looking to borrow $1,600 for a motorcycle, Louis says, "Sorry, pal. I have to charge you 37.5% interest because, you know, motorcycles are dangerous, and you might wreck it before you pay me back."

3. Sammy, like Manny, knows how to find 10% and 1% of any number, so he quickly calculates the interest he will pay. Show the numbers he might record to figure his interest payment.

4. If he wanted to check his numbers, how might Sammy figure the interest amount another way?

Activity 3: Decimal Equivalents

1. 100-Block Grid

   a. Shade $\frac{1}{8}$ of this grid. How many squares did you shade?

   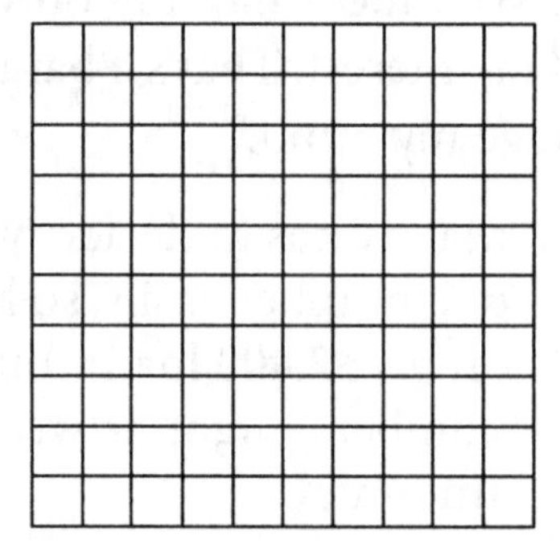

   b. Write another fraction that describes the part you shaded in the grid above.

   c. Write a percent that describes the part you shaded in the grid above.

   d. Write a decimal that describes the part you shaded in the grid above.

2. 1,000-Block Grid

   a. Make a rectangular grid with a 1,000 blocks by taping together 10 100-block grids. Shade $\frac{1}{8}$ of this grid.

   b. Write another fraction that describes the part you shaded in the 1,000-block grid.

   c. Write a percent that describes the part you shaded in the 1,000-block grid.

   d. Write a decimal that describes the part you shaded in the 1,000-block grid.

3. Predictions

a. How many blocks would be shaded if $\frac{5}{8}$ of the 1,000-block grid were covered?

b. What percent would that be? How do you know?

c. What decimal would you write for that fraction/percent? How do you know?

d. How many blocks would be shaded to show $\frac{1}{16}$ of the 1,000-block grid?

Practice: One-Eighth

Fill in the blanks to make the statements true.

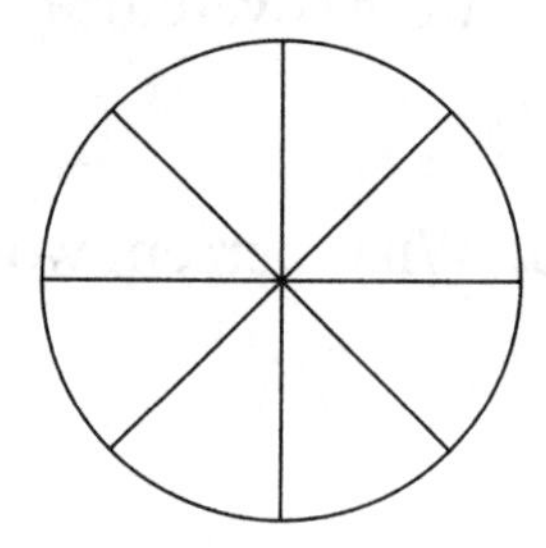

1. Half of a half is the same as______________________________.

2. Half of a fourth is the same as ______________________________.

3. Half of an eighth is the same as ______________________________.

4. To find half of any number, I can split that number into _____ groups. That is the same as dividing the number by _____.

5. To find a fourth of a number, I can split that number into _____ groups. That is the same as dividing the number by _____.

6. To find an eighth of a number, I can split that number into _____ groups. That is the same as dividing the number by _____.

7. If I want to find $\frac{3}{8}$ of a number, I can

____

____.

8. Another way that I can find $\frac{3}{8}$ is to

____

____.

Practice: Fair Shares of Candy Bars

The office candy dish holds 16 candy bars. Office workers share the bars equally.

1. If two people split the candy bars, what is each person's share?

   ____

   A fraction for each person's share is _____.

   A percent for each person's share is _____.

   Show on the grid how the bars are shared:

2. If four people split the candy bars, what is each person's share?

   ____

   A fraction for each person's share is _____.

   A percent for each person's share is _____.

   Show on the grid how the bars are shared:

3. If eight people split the candy bars, what is each person's share?

____

A fraction for each person's share is _____.

A percent for each person's share is _____.

Show on the grid how the bars are shared:

4. If 16 people split the candy bars, what is each person's share?

____

A fraction for each person's share is _____.

A percent for each person's share is _____.

Show on the grid how the bars are shared:

5. If 32 people split the candy bars, what is each person's share?

____

A fraction for each person's share is _____.

A percent for each person's share is _____.

Show on the grid how the bars are shared:

Practice: Designer Accessories

Sometimes combining percents gives a total that is simpler to understand. Sometimes it is easier to solve a problem with a fraction than a percent.

You run an accessory stall on weekends. Your stall has 96 square feet of display space that measures 16 feet by 6 feet.

You use

- $37\frac{1}{2}$% of the space to display purses;
- $37\frac{1}{2}$% of the space to display jewelry;
- 12.5% of the space to display hats and scarves; and
- 12.5% of the space to display belts.

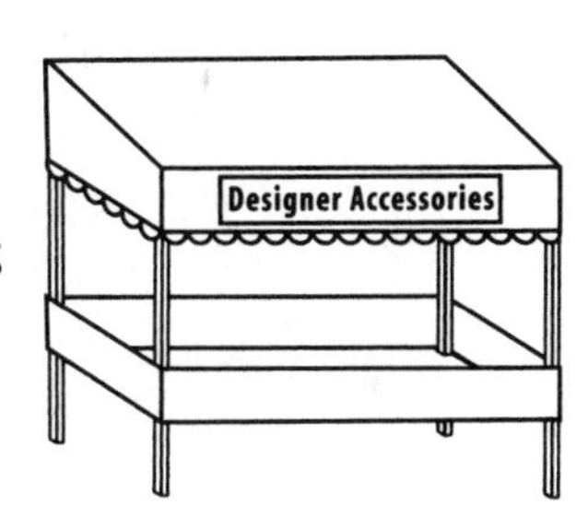

Use the grid to show how you could arrange your stall. Then answer the questions.

1. My designer accessory display plan:

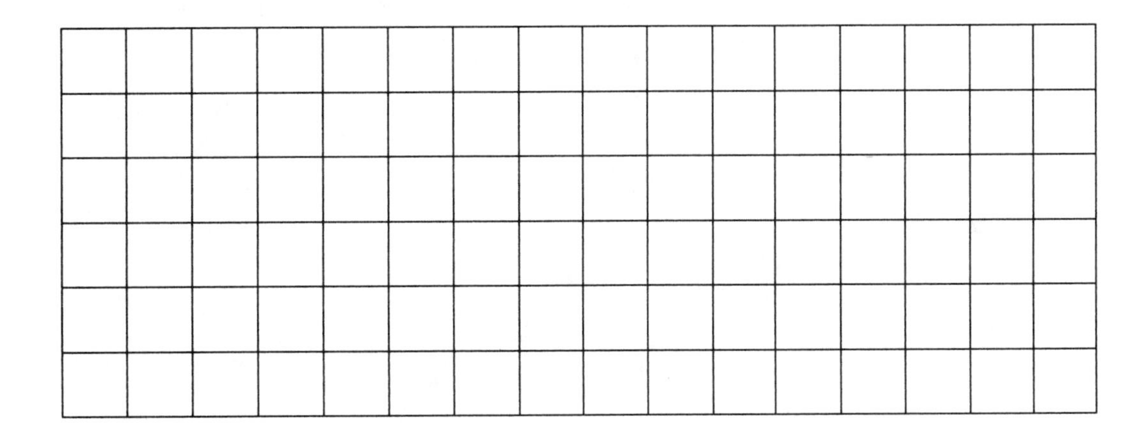

2. Key:

3. How many squares are needed to show the 12.5% used for displaying belts? Explain how you know with words, pictures, or numbers.

4. How many squares are needed to show the $37\frac{1}{2}$% used for displaying jewelry? Explain how you know with words, pictures, or numbers.

Practice: Round Up or Down?

Ted and Tina calculate their business expenses. They tell their bookkeeper to set aside 12.5% of sales for their retirement. However, they each round the percent to a whole number to estimate how much should be deposited in their retirement account.

Ted rounds down to 12%. Tina rounds up to 13%.

1. Complete the table to show what each one estimates for retirement deposits for six months. Show your work or describe your method for calculating percents.

Month	Total Sales	Ted's 12% Estimate for Retirement	Tina's 13% Estimate for Retirement
1	$ 1,000		
2	$ 800		
3	$ 6,500		
4	$ 4,000		
5	$ 1,600		
6	$10,000		

2. What total does Ted estimate will be in the retirement account after six months? What total does Tina estimate will be in the account?

3. After six months, what is the total amount the bookkeeper has deposited in Ted and Tina's retirement account? Explain how you know with pictures, words, and numbers.

4. How would *you* have estimated the amount in the retirement account? Why?

Extension: Precise Measurements

1. Locate and label each of the following measurements on the ruler:

$\frac{1}{8}''$ $3\frac{1}{2}''$ $\frac{3}{4}''$ $\frac{5}{8}''$ $1\frac{3}{8}''$ $\frac{1}{4}''$ $3\frac{7}{8}''$

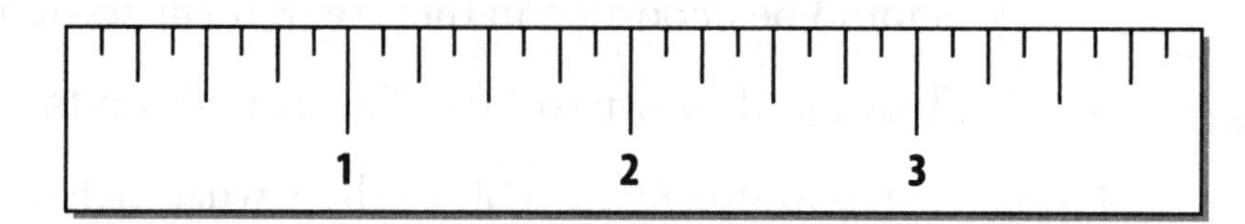

2. True or false?

a. $\frac{1}{8}'' > \frac{1}{4}''$

b. $\frac{3}{8}'' > \frac{1}{4}''$

c. $\frac{7}{8}'' < \frac{1}{2}''$

d. $\frac{5}{8}'' < \frac{1}{2}''$

e. $\frac{3}{8}'' < \frac{1}{2}''$

f. $\frac{7}{8}'' < \frac{3}{4}''$

Test Practice

1. Which of the following does *not* show $\frac{1}{8}$?

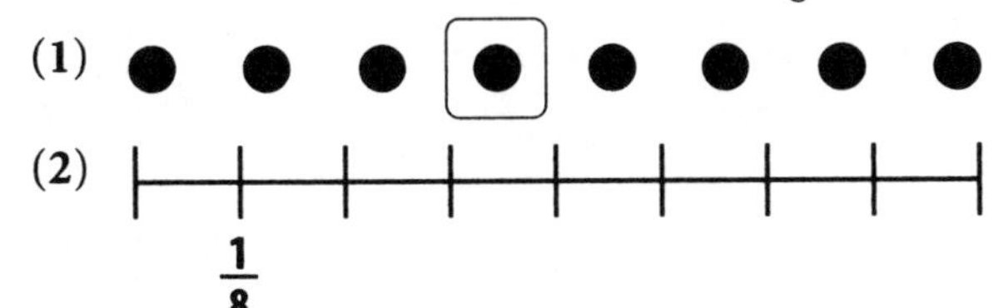

(1)

(2)

(3)

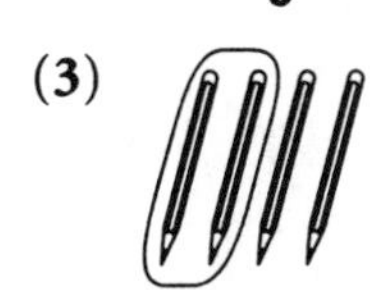

(4) $\frac{125}{1{,}000}$

(5) 0.125

2. José ate 16 cookies at the holiday party. His sister, Marcela, ate only two cookies because she is watching her weight. Marcela ate

(1) $\frac{1}{16}$ as many cookies as José.

(2) $\frac{1}{8}$ as many cookies as José.

(3) $\frac{14}{16}$ as many cookies as José.

(4) $\frac{1}{4}$ as many cookies as José.

(5) $\frac{1}{2}$ as many cookies as José.

3. Ellen pays her mother a fraction of the rent each month. If Ellen pays \$250 and the total rent is \$2,000, what fraction of the rent does Ellen's mother pay?

(1) $\frac{1}{8}$

(2) $\frac{1}{4}$

(3) $\frac{3}{8}$

(4) $\frac{5}{8}$

(5) $\frac{7}{8}$

4. The following fractions need to be ordered from least to greatest. What is the proper order for these fractions?

$\frac{3}{4}$ $\frac{1}{4}$ $\frac{1}{8}$ $\frac{3}{8}$ $\frac{7}{8}$

(1) $\frac{1}{4}, \frac{1}{8}, \frac{3}{8}, \frac{3}{4}, \frac{7}{8}$

(2) $\frac{1}{4}, \frac{1}{8}, \frac{3}{4}, \frac{3}{8}, \frac{7}{8}$

(3) $\frac{1}{8}, \frac{1}{4}, \frac{3}{8}, \frac{3}{4}, \frac{7}{8}$

(4) $\frac{1}{8}, \frac{1}{4}, \frac{3}{4}, \frac{3}{8}, \frac{7}{8}$

(5) $\frac{1}{8}, \frac{1}{4}, \frac{3}{8}, \frac{7}{8}, \frac{3}{4}$

5. Charlie is health conscious. He reads nutrition labels carefully. Eating an M's Dark Chocolate serving provides close to $\frac{1}{8}$ of the daily value for which nutrient?

M's Dark Chocolates

Nutrition Facts
Serving Size 6 pieces (42g)
Servings Per Container about 7

Amount Per Serving	
Calories 220	
Calories from Fat 120	
	% Daily Value
Total Fat 13g	**20%**
Saturated Fat 8g	**40%**
Polyunsaturated Fat	
Monosaturated Fat	
Cholesterol 5mg	**2%**
Sodium 0mg	**0%**
Total Carbohydrate 26g	**9%**
Dietary Fiber 3g	**12%**
Sugars 21g	
Protein 2g	

(1) Sodium

(2) Cholesterol

(3) Dietary fiber

(4) Saturated fat

(5) Protein

6. Alberto started a new job. He made \$16,000 a year in his old job. He received a 37.5% raise in his new job. How much will he make in a year at the new job?

LESSON 6

Give Me a Third

How can you find a third?

In this lesson, you will explore **thirds** and their percent equivalents. The fraction $\frac{1}{3}$ tells you two things right away: The total, or whole, is divided into three parts, and each of those parts is less than one-half.

Activity 1: One-Third as a Percent

1. Shade the grid to show $\frac{1}{3}$ of it.

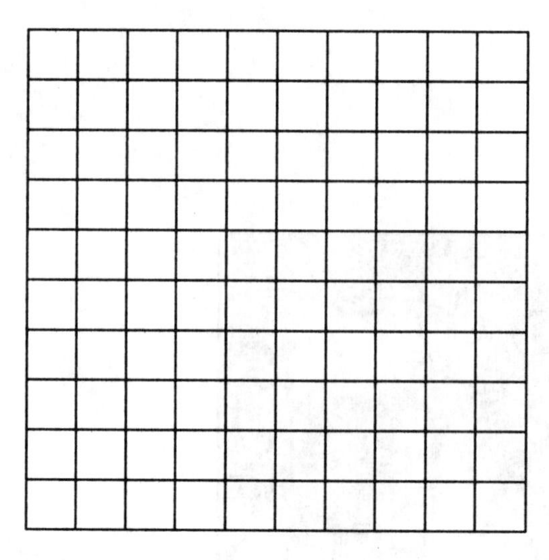

2. How many squares did you shade to show one-third of the grid?

3. How would you name $\frac{1}{3}$ as a percent?

4. Circle the squares to show one-quarter of the grid.

5. Which is bigger, one-fourth or one-third?

Activity 2: Stations

We often have a sense of a fractional amount just by looking. For example, you may have an idea about how much of a book you have read or how much of a glass of water you drank.

At each station:

1. Estimate, then calculate, measure, or collect data to determine the *closest* fraction and percent represented in each situation.

   Choose from benchmark fractions: $\frac{1}{10}$, $\frac{1}{4}$, $\frac{1}{3}$, $\frac{1}{2}$, $\frac{2}{3}$, or $\frac{3}{4}$.

2. Explain with words, pictures, or numbers how you reached your conclusion.

Question	Closest Fraction	Closest Percent	Explanation of Your Reasoning
How full is the container?			
What part of the book has been read?			
What part of the class has ever worked in the food industry?			
What portion of the classes have you attended?			
What fraction of the paper clips is left in the box? What percent?			
How much money did I have at first?			

Activity 3: News Report

Each of the following problems contains a statistic. Write a news report to announce the statistic.

- Invent numbers to give the statistic meaning.
- Use fractions and percents in your report.
- Prepare a visual to show the class. For example, make a number line, circle graph, or shaded grid, and label it to show the breakdown of fractions or percents.

The first problem is started for you.

1. One-third of all e-mail is spam (or junk mail).

A new study found that one-third of all e-mail is spam. This means that when a worker receives 50 e-mails a day, ...

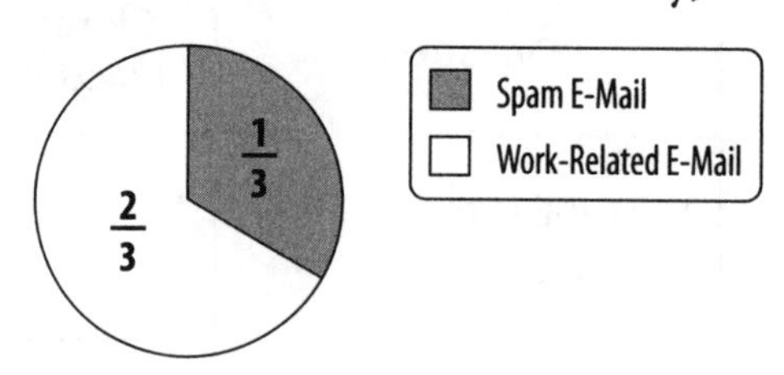

2. One-third of all U.S. adults have hypertension.

3. One of every three lakes in the United States and nearly one-third of the nation's rivers are so polluted that people should limit or avoid eating fish caught in them.

4. One-third of the world's urban population lives in slums.

5. Some middle-class workers are now spending two-thirds of their monthly take-home pay on rent.

Practice: Show Me $\frac{2}{3}$!

Show the portion that equals $\frac{2}{3}$ of each of the following shapes or sets of objects. Clearly mark the thirds. Then fill in the blanks.

1.

a. The total number of pieces (the whole) is ______.

b. The number of pieces shaded (the part) is ______.

c. The fraction is ______.

2.

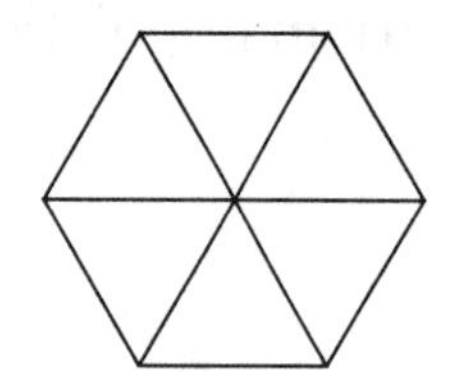

a. The total number of pieces (the whole) is ______.

b. The number of pieces shaded (the part) is ______.

c. The fraction is ______.

3.

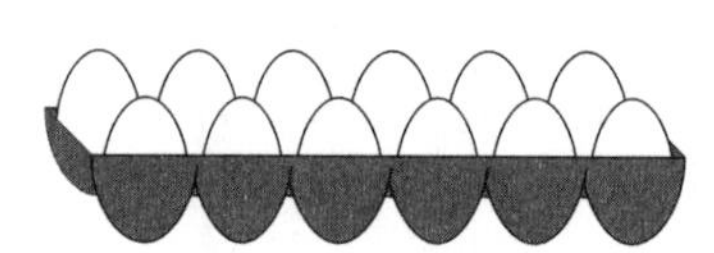

a. The total number of pieces (the whole) is ______.

b. The number of pieces shaded (the part) is ______.

c. The fraction is ______.

4. Mark $\frac{2}{3}$ of the distance from zero on each of the following number-line segments, and label the value.

a.

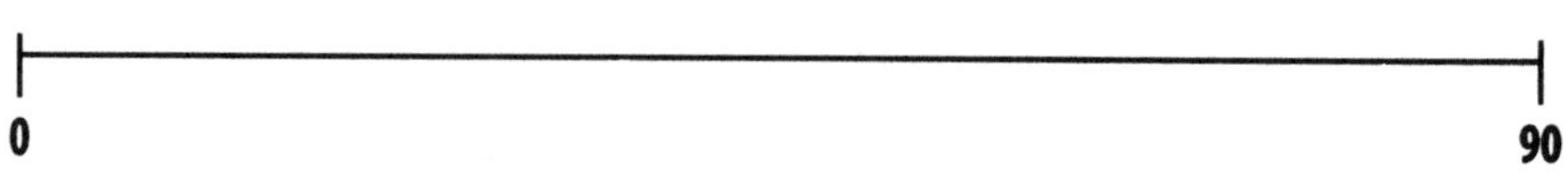

b.

0 45

c.

0 15

Practice: Comparing Tips

Three housemaids compared tips. Each time Michelle had received the most generous tips.

Shade each bar to show the amounts based on your estimates. Then use math to check your estimates. If you were more than $10 off, draw a new bar with the right amounts shown.

1. In a full shift, Michelle earned $100 in tips. Susan earned $\frac{2}{3}$ of Michelle's amount. Cara earned $\frac{1}{3}$ of Susan's tips.

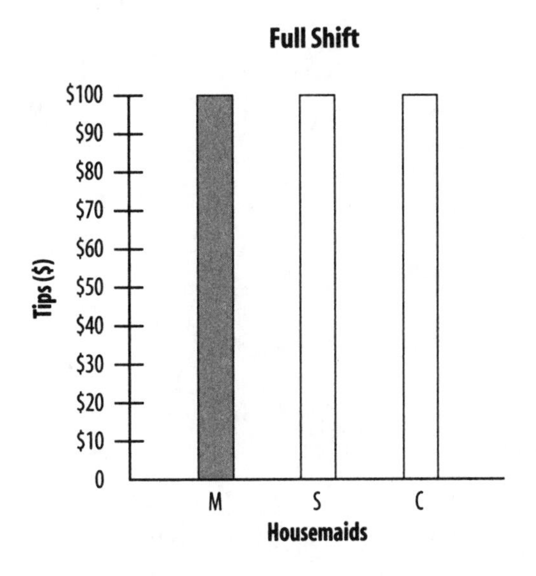

2. On a shorter shift, Michelle earned $50. Susan earned $\frac{2}{3}$ of Michelle's amounts. Cara earned $\frac{1}{3}$ of Susan's tips.

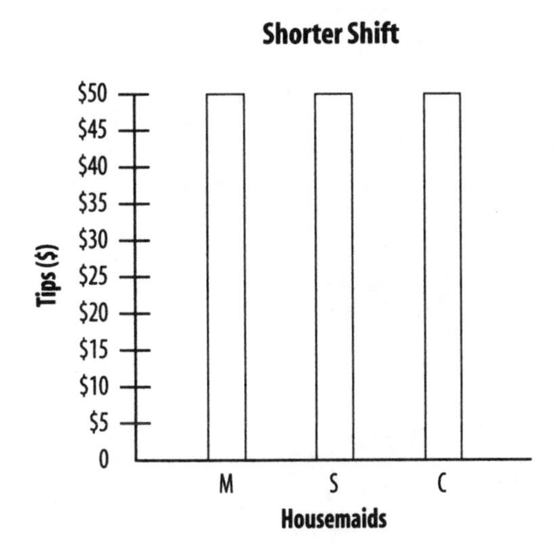

3. Michelle earned $14 in the first hour of her shift. Susan earned $\frac{2}{3}$ of that. Cara earned $\frac{1}{3}$ of Susan's tips.

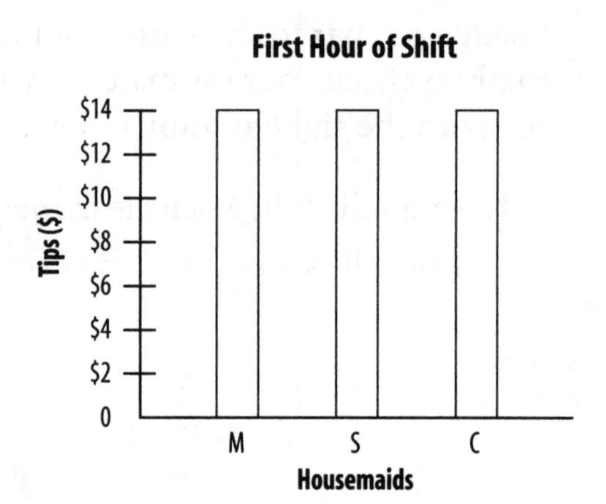

Practice: Better Deal?

Part 1: The Facts

Before you start, make a cheat sheet for yourself.

1. To find $\frac{1}{3}$ ______________________________

   ______________________________.

2. To find $\frac{2}{3}$ ______________________________

   ______________________________.

3. To find 30% ______________________________

   ______________________________.

4. In percents, $\frac{1}{3}$ = ____________.

5. Choose the correct statement.

   a. $30\% < \frac{1}{3}$

   b. $30\% = \frac{1}{3}$

   c. $30\% > \frac{1}{3}$

Part 2: Which Deal is Better?

Compare the sales prices. For each item, circle the sales price your instinct tells you is lower. Then do the math to find out if you were correct. Show your work.

1. Jacket

At Barkers:

At Elsa's:

2. Set of Kitchen Knives

At Barkers:

At Elsa's:

3. Portable CD Player

At Barkers:

At Elsa's:

4. Running Shoes

At Barkers:

At Elsa's:

5. Dollhouse

At Barkers:

At Elsa's:

Practice: How Much, How Far?

1. Val and Solé plan to join the AIDS Walk-a-Thon. Val is responsible for walking one-third the distance and raising one-third of the pledges. Solé is responsible for walking two-thirds the distance and raising two-thirds of the pledges.

   a. The AIDS Walk-a-Thon covers 24 miles. Val walks five miles and stops. Is this fair? Why or why not? Use a diagram or number line to support your thinking.

   b. Val raised $1,000, which was 33.33% of the money. If Solé raised her share, this means Solé raised _______%, or $_______.

   c. Together the two raised _______% or $_______.

   d. How do you know?

2. Dana and Martin walk 21 blocks to the video store every Friday. They always stop two-thirds of the way there to buy snacks.

   a. At what block do they stop? _____
   Show with a diagram or number line how you know.

   b. What fraction of their walk remains?

   c. Is that fraction more or less than one-fourth? Explain.

Extension: Budgeting

Use tenths, thirds, halves, and quarters to figure out monthly budgets. Then list the percent equivalents for each fraction. Check your math by finding the total.

1. Family size: two—a single mother with a school-age son
   Monthly income: $1,200

Expenses	Dollar Amount	Fraction of Total Monthly Income	Percent of Total Monthly Income
Food and drugs			
Rent			
Utilities			
Clothing/school supplies			
Transportation			
Other			
Total			

2. Create your own example.

Family size:

Monthly income:

Expenses	Dollar Amount	Fraction of Total Monthly Income	Percent of Total Monthly Income
Food and drugs			
Rent			
Utilities			
Clothing/school supplies			
Transportation			
Other			
Total			

Test Practice

Problems 1 and 2 refer to the following statistic: During a recent hurricane, one-third of all homes on the island were damaged by wind. The count of damaged homes is 230.

1. How many homes were not damaged by wind?

(1) 230

(2) 460

(3) 490

(4) 630

(5) 690

2. Of the damaged homes, about 30% were declared completely destroyed. How many houses were considered a total loss?

(1) Close to 10

(2) Close to 25

(3) Close to 50

(4) Close to 70

(5) Close to 100

3. Tripling a recipe that calls for one-third cup of oil requires how many cups of oil?

(1) $\frac{1}{3}$ cup of oil

(2) $\frac{2}{3}$ cup of oil

(3) 1 cup of oil

(4) 3 cups of oil

(5) 9 cups of oil

4. A small store tracked the payment type chosen by its customers during one day. According to the graph, about what fraction of purchases were made with credit cards?

ATM Debit	Credit	Cash
卌 卌 卌 ///	卌 卌 卌 卌 卌 卌 卌 卌 卌 卌 卌 卌 卌 /	卌 卌 卌

(1) $\frac{1}{4}$

(2) $\frac{1}{3}$

(3) $\frac{4}{10}$

(4) $\frac{1}{2}$

(5) $\frac{2}{3}$

5. The credit card company charges the store 2% for its service. On sales of $2,500, the store would pay how much?

(1) $2

(2) $15

(3) $20

(4) $25

(5) $50

6. The budget for major street work in a business area includes "mitigation" money to compensate business owners for lost revenue due to construction. The reasoning is that stores will lose money because not as many people will shop if the street is a mess. Refer to the graph to answer the question: How much money is allocated for mitigation in the Public Works project?

$90,000 Public Works Project

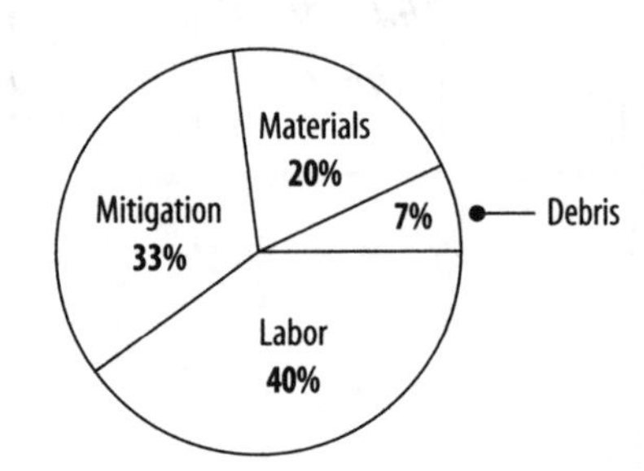

LESSON 7

Order, Choose, and Change

What's the order?

In this lesson, you will consolidate what you learned in this unit. Your knowledge about familiar fractions and decimals and their relation to percents will all come into play as you solve the problems in this lesson.

You will compare benchmarks and order them by size. You will complete word problems and solve them in ways you find comfortable, and you will determine a whole, given a part.

Activity 1: Ordering Benchmarks

My group is working with (circle one): Decimals Fractions

Order your cards in size from least to greatest.

1. Write the order your group agreed on below:

2. After the class agrees on an order for the fractions, decimals, and percents, complete the following table ordering from least to greatest.

Percents	Fractions	Decimals

Loaves of Bread

Five hungry friends went to the baker, and each bought a mini-loaf of bread. *Each loaf was the same size.*

Rich cut his loaf into four slices and ate three of them.

Jewel cut her loaf into five slices and ate four.

Bea cut her loaf into 10 slices and ate nine.

Chen cut his loaf into 20 slices and ate 17.

Luis cut his loaf into 16 slices and ate 14.

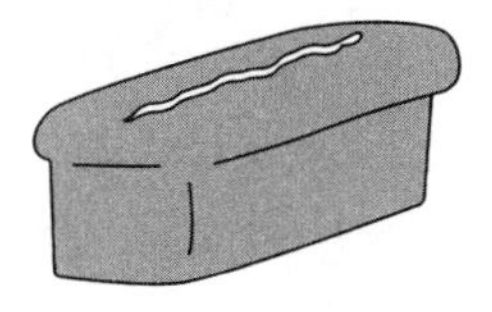

3. Who ate the most bread? How do you know?

4. Who had the most bread left over? How do you know?

5. If Chen had eaten one more slice, would he have eaten the most bread? What if he had eaten two more slices?

6. Tasha joined the group with her own mini-loaf of bread—the same size loaf that the others had bought. She said, "I cut my loaf into more slices than any of you, and I ate more slices than any of you. But I have more of my loaf left than any of you."

Could Tasha have been telling the truth? Why or why not? Give examples to show your reasoning.

Activity 2: Choosing the Way

Fill in the blanks to create a problem, and then explain how you solved it.

- Choose a number that makes sense for the first blank.
- Use a fraction, decimal, or percent from the table completed in *Activity 1*, page 90, to fill in the second blank. Use at least one fraction and one percent in the first five problems.
- Write the answer to the problem in the third blank.
- Use words and numbers to explain how you solved the problem in the fourth blank.

1. Jack and Jill decide to share the cost of a new car. The car cost $________. They plan to borrow some of the money, but they paid _____ of the cost as a down payment. This means they gave the car dealer $_______ and borrowed the rest. They figured out how much to give the car dealer by ______________________________

____

____

____.

2. Each day _____ people shop at the local supermarket. Of those shoppers, _____ are buying food only for themselves. That means that _____ people each day buy food for themselves alone. The store manager figured this out by ___________________________

____

____

____.

3. Kevin drove _____ miles to reach his parents' home. He stopped after he completed _____ of his trip. That meant he had traveled _____ miles. Kevin knew this because he ______________________

____

____

____.

4. Colin visited a farm with _____ animals. He counted how many were brown and how many were black. He told his sister that _____ of the animals were brown. She then figured that _____ of the cows, dogs, horses, and hens were brown. Colin said she was right. She knew she was right because she had ______________________

____

____

____.

5. Vinnie is an excellent typist. He types ____ words per minute. He makes very few errors. In fact, only _____ of his words contain "typos," or are misspelled. This means that each minute he makes only _____ errors.

   Vinnie figured out his fraction/percent of errors by ____________

____

____

____.

6. Create your own word problem:

7. Did you use *fractions*, *decimals*, or *percents* most often to *solve* Problems 1–6? (Circle one.) Why?

Activity 3: What Was the Whole?

Read each problem, and use what you know about fractions and percents to find the solution.

Use pictures, words, or numbers to support your solutions.

1. The local radio station played 24 songs by women yesterday. How many songs did they play if female performers represented

   a. $\frac{1}{3}$ of the total? ______

   b. $\frac{3}{8}$ of the total? ______

   c. 40% of the total? ______

   d. 5% of the total? ______

2. Our office party ended, and six people stayed to clean up. What fraction (or percent) of the total number of people who attended the party could that have been?

   Find at least five different possibilities. For each one, write down the fraction and the total.

Practice: More Word Problems

Fill in the blanks to create a problem, and then explain how you solved it.

- Choose a number that makes sense for the first blank.
- Use a fraction, decimal, or percent from the table completed in *Activity 1*, page 90, to fill in the second blank. Use at least one fraction and one percent in the four problems.
- Write the answer to the problem in the third blank.
- Use words and numbers to explain how you solved the problem in the fourth blank.

1. Yuri usually pays about _____ for utilities each month (gas, phone, computer cable, electricity, etc.). One month his bills skyrocketed. They increased by _____. Yuri had to pay an extra _____. Yuri figured out how much his bills had increased by ______________

____

____

____.

2. Janice eats _____ calories a day. She knows that _____ of her calories come from the fat she eats. This means Janice eats _____ calories of fat each day. She figured this out by ______________

____

____

____.

3. Mona signed a contract to buy a condo with the money she saved. The condo cost _____. Mona will take out a mortgage to pay for most of the condo, but she intends to pay for _____ of the condo in cash. She will pay _____. Mona calculated how much she would pay by ______________________________

____

____

____.

4. Danny has a CD collection with _____ CD's. _____ of his collection is devoted to hip-hop music. That means he has _____ hip-hop CD's. He figured out the fraction/percent of hip-hop CD's by

____

____

____

____.

Practice: A Big Brownie

Three friends shared a big brownie. Each person ate a different fraction of the brownie. Together they ate the whole brownie.

1. What fraction could each person have eaten? Complete a row of the chart for each way they might have shared the brownie. Try to find at least two ways that use fractions with different denominators.

What fraction did each person eat? Write the fractions in order of size.	How do you know that all the fractions together equal 1? Use pictures, numbers, and/or words to explain your reasoning.

2. Who ate the most? Who ate the least? Order the fractions of brownie you listed in the table from the smallest to the largest piece.

Practice: New Student Housing

5% 10% 12.5% 15% 20% 25% 37.5% 50% 60%

Choose from the percents above to design your own plan of a dorm space with three bedrooms and a common area. Use four different percents. Label or color each area and fill out the key to match.

_______% is ____.

_______% is ____.

_______% is ____.

_______% is ____.

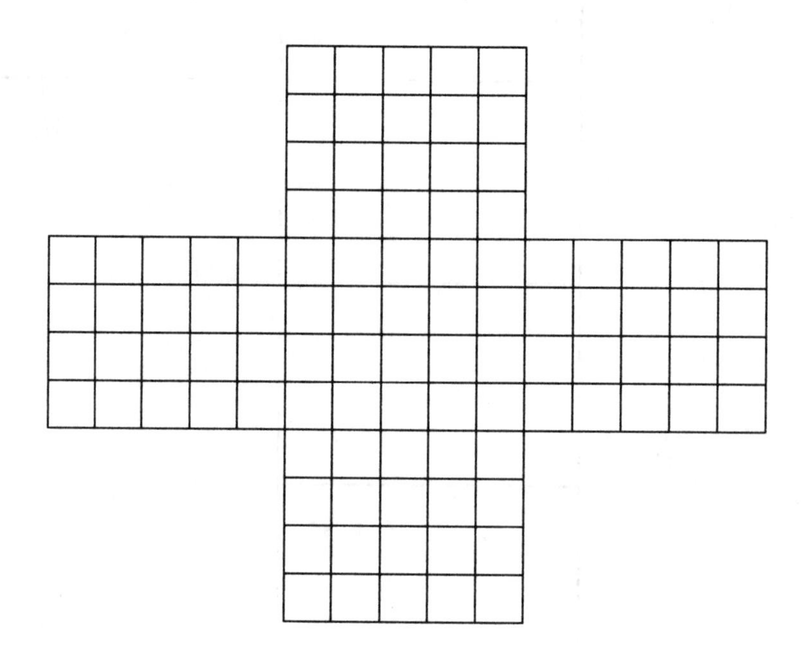

Key:

Practice: Two Truths and a Lie

For each problem, you will see three ways of finding its solution. Two of the ways are correct ways to solve the problem; one is false.

- Mark the true statements "T" and the false statements "F."
- Correct the false statements to make them true.

1. Find 60% of the money raised when $800 is raised.

   ____ **a.** Find 0.6 times $800.

   ____ **b.** Find 10% of $800 and multiply that by 60.

   ____ **c.** Find $\frac{3}{5}$ of $800.

2. Find 15% of the voters in a town with 1,200 registered voters.

   ____ **a.** Find 10% of 1,200. Find half of 10% of 1,200. Add the two numbers.

   ____ **b.** Find $\frac{1}{4}$ of 1,200. Find $\frac{1}{10}$ of 1,200. Subtract the $\frac{1}{10}$ number from the $\frac{1}{4}$ number.

   ____ **c.** Find 1.5 times 1,200.

3. Find 12.5% interest on a loan of $480.

   ____ **a.** Find 10%, 1%, and 0.5% of $480.

   ____ **b.** Find $\frac{1}{8}$ of $480.

   ____ **c.** Find 0.125 times $480, and then add those amounts.

4. Find $66\frac{2}{3}$% of miles traveled on a 150 mile trip.

   ____ **a.** Find $\frac{2}{3}$ of 150.

   ____ **b.** Find 0.666 times 150.

   ____ **c.** Find 150 ÷ 2, and then triple the result.

5. Find 80% of tenants in a building with 620 residents.

   ____ **a.** Find 20% of 620 and subtract that number from 620.

   ____ **b.** Find $\frac{1}{8}$ of 620.

   ____ **c.** Find 0.8 times 620.

Practice: Comparing Schedules

Nina and Fanny work as waitresses. They job share. Look at their schedules and answer the questions that follow.

Nina and Fanny's Schedules
Week of August 4

Nina	Fanny
Monday Night	**Monday Night**
7–8 p.m.—Bike riding	7–9 p.m.—Work
Tuesday Night	**Tuesday Night**
6–10 p.m.—Work	
Wednesday Night	**Wednesday Night**
5–8 p.m.—Volunteer	4–10 p.m.—GED class
Thursday Night	**Thursday Night**
6–10 p.m.—Work	4–6 p.m.—Work
Friday Night	**Friday Night**
5–9 p.m.—Work	5–8 p.m.—Bowling
Saturday Night	**Saturday Night**
6–10 p.m.—Ballroom dancing	5–7 p.m.—Work

1. Nina divides their tip money for the week. She keeps $\frac{2}{3}$ of the money and gives Fanny $\frac{1}{3}$. Do you think the way Nina and Fanny divide the tips is fair? Why or why not?

2. Write a statement that compares Nina's time spent furthering her education (her GED class) and Fanny's time spent volunteering. Use a fraction or percent in your statement. Show with pictures, words, or numbers why your statement is true.

Extension: The Whole Is . . .

Explain your answer in each problem.

1. For his construction work, Alberto asks to be paid one-third of the total payment at the end of the first week of work. In his last job, he received $3,000 at the end of the first week. The whole job cost ________.

2. Forty-five donuts were sold at the donut shop in the first hour on Tuesday. That was 75% of the donuts baked early that morning. The number of donuts baked was ________.

3. Two-thirds of the town votes were needed to pass the bill to fund a new library, and we got it! There were 700 "yes" votes. What was the total number of votes?

4. Christina, who loves music, is happy because $\frac{5}{8}$ of the radio stations in her area play music that she likes. Fifteen stations play Christina's favorite music. How many radio stations are there in her area?

Test Practice

1. If the following fractions were listed from least to greatest, which fraction would be next to last?

   $\frac{1}{20}$ $\frac{1}{100}$ $\frac{3}{4}$ $\frac{1}{3}$ $\frac{1}{2}$ $\frac{2}{3}$ $\frac{3}{5}$

   (1) $\frac{1}{100}$

   (2) $\frac{1}{2}$

   (3) $\frac{3}{5}$

   (4) $\frac{2}{3}$

   (5) $\frac{3}{4}$

2. If the following decimals were listed from least to greatest, which one would be listed second?

   0.1 1.0 0.25 0.333 0.01 0.125 0.05

   (1) 0.1

   (2) 1.0

   (3) 0.125

   (4) 0.01

   (5) 0.05

3. Which of the following expressions shows how to find 17% of a 200-volume book collection?

   (1) 1.70(200)

   (2) 17(200)

   (3) 200 ÷ 17

   (4) 200 ÷ 10 + 7(200 ÷ 100)

   (5) 200 ÷ (100 x 17)

4. Eight people stayed to talk to the guitarist after the show. This represented 12.5% of the audience. How many people were in the audience?

   (1) 8

   (2) 24

   (3) 64

   (4) 80

   (5) 100

5. Last year Soren earned $2,500 toward his college tuition. The college is asking him to contribute 12.5% more this year. How much money will Soren need to make for college this year?

   (1) $2,512.50

   (2) $2,625.00

   (3) $2,775.00

   (4) $2,812.50

   (5) $3,124.00

6. Write a decimal number that is more than $\frac{1}{3}$ and less than $\frac{3}{4}$.

LESSON **8**

Where's the Fat?

How can you tell if a food is high in fat?

Watching what you eat and learning how to read and interpret nutrition labels is important for your overall health. Reading labels is one way to become informed about the nutritional content of foods.

In this lesson, you will read nutrition labels and compare fat **calories** with total calories per serving to determine the percent of calories from fat in various foods.

Using what you know about fractions, decimals, and percents, you will determine if a food is nonfat, low fat, medium fat, or high fat.

Activity 1: How Much Fat?

Part 1: Nutrition Facts

Walnuts
Nutrition Facts
Serving Size 1/4 cup (30g)
Servings Per Container about 45
Amount Per Serving
Calories 210
Calories from Fat 180

1. How many calories are there in a serving of walnuts? _______
2. How many of those calories are from fat? ________
3. What fraction of the calories comes from fat? ________
4. Is that fraction less than, equal to, or more than $\frac{1}{2}$? _______ Explain.

5. Is the fraction more than $\frac{3}{4}$? ____ Explain.

Part 2: Nutrition Labels

The 2005 edition of the *Dietary Guidelines for Americans*, published by the Department of Health and Human Services (HHS) and the Department of Agriculture (USDA), gives advice about healthy dietary habits. One of the recommendations is as follows:

> "Keep total fat intake between 20 to 35 percent of calories, with most fats coming from sources of polyunsaturated and monounsaturated fatty acids, such as fish, nuts, and vegetable oils."

Although the recommended 20 to 35 percent of calories from fat is in reference to the *daily* total number of calories in a healthy diet, you might use these numbers as a guide when you look at individual food labels.

Look closely at the nutrition labels to determine the fat content.

1. Sort the labels into four groups: Nonfat (0%), low fat (less than 20% fat calories, but more then 0%), medium fat (20%–35%), or high fat (more than 35% calories from fat).

2. Complete the following chart with information from at least one of each type of fat category.

Food	Total Calories per Serving	Calories from Fat	Estimated % Fat (Ballpark Guess!)	Benchmark Fraction	Percent of Fat Calories	High Fat, Medium Fat, Low Fat, or Nonfat

3. What food had the highest percent of fat content? How did you know?

4. Did the percent of fat content of any of the foods surprise you? Why?

5. What method did you use to determine the percent of fat? Show your work.

6. How can you check your work? Show another method.

Activity 2: Munching on Snacks

1. What is the percent of fat calories in these two snacks?

Snack 1

Food	Calories per Serving	Calories *from Fat* per serving
Orange juice	120	0
Chocolate chip cookie	120	60

a. Percent of fat calories in Snack 1:

Snack 2

Food	Calories per Serving	Calories *from Fat* per serving
Milk	150	70
Chocolate chip cookie	120	60

b. Percent of fat calories in Snack 2:

Isadore's Market

Isadore's Market sells four kinds of lemon cookies.

Lucy's Lemon Cookies

Nutrition Facts
Serving Size 2 cookies
Servings Per Container about 20

Amount Per Serving
Calories 120
Calories from Fat 60

Lite Lemon Cookies

Nutrition Facts
Serving Size 2 cookies
Servings Per Container about 30

Amount Per Serving
Calories 200
Calories from Fat 50

Diet Joy Lemon Cookies

Nutrition Facts
Serving Size 2 cookies
Servings Per Container about 15

Amount Per Serving
Calories 90
Calories from Fat 60

Little's Tiny Lemon Cookies

Nutrition Facts
Serving Size 8 cookies
Servings Per Container about 20

Amount Per Serving
Calories 100
Calories from Fat 50

Decide which cookie is the best nutritionally. Take into account the calories from fat content, but use other information as well. Write a radio ad to convince the public that your cookie is nutritionally the best choice.

Refer to the lemon cookie labels from Isadore's Market to answer Problems 1–6.

1. Which cookies have the highest percent of fat calories?

2. Which cookies have the lowest percent of fat calories?

3. Which two kinds of cookies have the same percent of fat calories?

4. Inez was so hungry that she ate two Lite Lemon Cookies and eight Little's Tiny Lemon Cookies (one serving of each kind of cookie). What percent of fat calories was her snack?

5. Felipe also ate two kinds of cookies for a snack (one serving of each). His snack was more than 50% fat. What two kinds of cookies might he have eaten?

   Find at least two different answers. For each answer, explain how you know the fat content is more than 50%.

6. Mahalia decided to eat only foods that have one-third or fewer calories from fat. She says that she can eat any of the lemon cookies sold at Isadore's Market as long as she only eats half a serving. Do you agree? Why or why not?

Practice: Counting Calories

Ana: "I'm supposed to be on a low-fat diet, but my breakfast was 100% fat!"

Tina: "What happened?"

Ana: "I had peanut butter and toast. The peanut butter was 135 fat calories out of 180 calories in all. That's 75% fat. The bread was 15 fat calories out of 60 calories in all—that's 25%. Twenty-five percent and 75% is 100%."

Tina: "You're always exaggerating! The fat is high, but not that high. Your breakfast was less than 75% fat."

Who do you agree with? Why? Use math in your explanation.

Practice: What Percent?

Twenty students are in the learning center math class, 15 women and five men.

1. What percent of the students in the class are women?

2. What percent of the students in the class are men?

3. Nine of the women are over 30 years old. What percent of the women is that?

4. What percent of the whole class is made up of women over 30?

5. Four of the men in the class are over 30 years old. What percent of the men is that?

6. What percent of the whole class is made up of men over 30?

Practice: What Fraction?

1. There are two math classes that meet at the same time on Tuesday afternoons. One class has 14 students, of which six are male, and the other has 12 students, of which only two are male.

   a. What fraction of the first class is male?

   b. What fraction of the second class is male?

   c. What fraction of both classes is male?

   d. Show your work.

   e. Is that fraction more or less than $\frac{1}{3}$? Explain.

2. Saul sells fish on Wednesdays, when he goes fishing. Last week he caught 16 sea bass and sold 12. Two weeks ago, he caught 24 and sold 20.

   a. What fraction of the sea bass did he sell last week?

   b. What fraction of the fish did he sell two weeks ago?

   c. What fraction of the fish he caught on both occasions did he sell? Show your work.

   d. Is that fraction more or less than $\frac{3}{4}$? Explain.

Practice: Rounding to the Hundredths Place

When you use a calculator, the answer can have many decimal places. To make the numbers more manageable, round them to two decimal places.

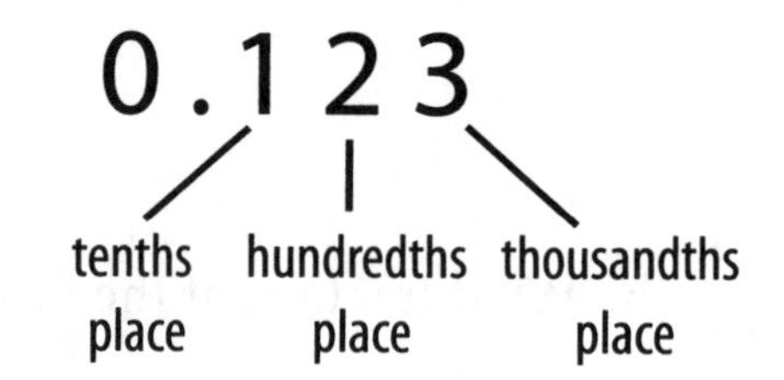

If the digit in the thousandths place is 4 or less, round down: Drop it and use only the tenths and hundredths digits.

Examples:

0.164	rounds to	0.16
0.8739	rounds to	0.87
0.22222...	rounds to	0.22

If the digit in the thousandths place is 5 or more, round up: Increase the digit in the hundredths place by 1.

Examples:

.125	rounds to	.13
.3071	rounds to	.31
.77777...	rounds to	.78
.297	rounds to	.30

A "9" in the hundredths place changes to "0," and the digit in the tenths place increases by 1.

Rounding to the Hundredths Place *(cont'd)*

Round each decimal to the hundredths place.

Decimal Number	Rounded
0.612	
0.105	
0.77777...	
0.095	
0.3005	
0.396	
0.999	
1.334	
1.909	
8.998	

Extension: What Is the Fat Content in My Food?

Keep track of the foods you eat for one meal.

Make a list of the foods.

When available, look at the nutrition labels and write down the total calories and calories from fat for your portion.

Calculate the total percent of fat calories in the meal.

Answer the question: What is the fat content in my food?

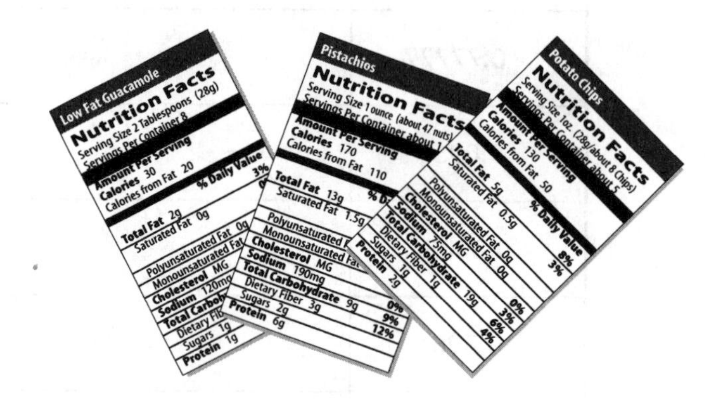

Test Practice

1. Mariza had a cup of soup for lunch. The total calories in that cup were 100, and the calories from fat were 25. What was the percent of fat calories?

   (1) 5%

   (2) 10%

   (3) 15%

   (4) 25%

   (5) 75%

2. Maia had a cup of miso soup for lunch. The total calories in her cup were 30, and the calories from fat were 10. What was the percent of fat calories?

   (1) 10%

   (2) 30%

   (3) 33%

   (4) 40%

   (5) 66%

3. Jon and his friends are munching on yogurt potato chips and pistachios. The yogurt potato chips have 130 calories per serving, 50 from fat. The pistachios have 170 calories per serving, 110 from fat. Jon calculates he ate one serving of each, yogurt potato chips and pistachios. Which best describes the total percent of fat calories in Jon's snack?

   (1) About $33\frac{1}{3}$%

   (2) A little less than 50%

   (3) Just over 50%

   (4) Just about 100%

   (5) Just about 160%

4. Zach, who was with Jon, ate two servings of pistachios and one serving of yogurt potato chips. Which best describes the total percent of fat calories in Zach's snack?

   (1) 30%

   (2) Just under 50%

   (3) Just over 50%

   (4) Almost 75%

   (5) Almost 100%

5. Rachel ate only half a serving of pistachios. The percent of fat calories for her snack is

   (1) The same as for one serving of pistachios.

   (2) Half as much as for one serving of pistachios.

   (3) Twice as much as for one serving of pistachios.

   (4) Four times as much as for one serving of pistachios.

   (5) One-fourth as much as for one serving of pistachios.

6. Luis has 24 pens on his desk. Three of his pens are red and six are blue. What fraction of his pens are red?

Closing the Unit: Put It Together

How well do you know your benchmarks?

In this unit you have expanded the collection of benchmark fractions from $\frac{1}{2}$, $\frac{1}{4}$, $\frac{3}{4}$, and $\frac{1}{10}$ to include $\frac{1}{100}$, $\frac{1}{3}$, $\frac{1}{8}$, and their multiples. You used these fractions and their equivalent decimals and percents to solve problems.

You found $\frac{1}{4}$ by taking half of $\frac{1}{2}$, and when you again split that number in half, you found $\frac{1}{8}$. You are now able to find halves of other fractions, for example, half of $\frac{1}{3}$ and half of $\frac{1}{8}$.

In this lesson, you will put it all together as you review the unit in preparation for your *Final Assessment.*

Activity 1: Review Session

Review what you learned in this unit.

- Go back to the practice pages in past lessons.
- Pick a page in each lesson.
- Cover up what you wrote on the page.
- Read the question only.
- Answer out loud or on paper.
- Reread your original answers to refresh your memory and check your work.

Determine whether a percent is less than, equal to, or more than 10%	*Lesson 1*
Given 10%, determine a whole	*Lesson 1*
Use percents and fractions to find 10% of a given amount	*Lesson 2*
Use 10% and 1% to calculate other percents	*Lessons 3 and 4*
Use percents and fractions to describe a markdown	*Opening, Lessons 4 and 5*
Find fraction, decimal, and percent equivalents for $\frac{1}{8}$	*Lesson 5*
Determine eighths of a given amount	*Lesson 5*
Find fraction, decimal, and percent equivalents for $\frac{1}{3}$	*Lesson 6*
Determine thirds of a given amount	*Lesson 6*
Order fractions, percents, and decimals	*Lesson 7*
Compare fractions with percents	*Lessons 7 and 8*
Find the combined percents of two or more quantities	*Lesson 8*

Making a portfolio of your best work will help you review the concepts and skills in this unit (see *Reflections* for *Closing the Unit: Put It Together*, page 134, for ideas).

Activity 2: Final Assessment

Complete the tasks on the *Final Assessment*. When you finish, compare your first Mind Map, page 2, with your Mind Map from the *Final Assessment*.

What do you notice?

VOCABULARY

Lesson	Terms, Symbols, Concepts	Definitions and Examples
Opening the Unit	decimal	
	denominator	
	fraction	
	numerator	
	part/whole	
	percent	
1	benchmark	
	estimate	
	rounding	
	ten percent, 10%	
2	multiples of 10%	
	25%	
	75%	

VOCABULARY *(continued)*

LESSON	TERMS, SYMBOLS, CONCEPTS	DEFINITIONS AND EXAMPLES
3	one percent, 1%	
4	gross pay	
	multiples of 1%	
	net pay	
5	one-eighth, $\frac{1}{8}$	
6	one-third, $\frac{1}{3}$	
8	calories	

REFLECTIONS

OPENING THE UNIT: Split It Up!

Find an item you bought on sale. Describe the markdown using fractions or percents. How does knowing benchmark percents help you find the markdown?

LESSON 1: Numbers in the News

What do you want to remember about 10%?

Where might you see 10%?

What do you picture when you hear "10%"?

LESSON 2: What Is Your Plan?

How can finding 10% of a number help you find 20% or 30% of a number?

Use the table below to keep track of fraction, decimal, and percent equivalents. As you learn more equivalents, add them to the table.

Fractions/Decimals/Percents Equivalents Chart

PERCENT TERMS	DECIMAL TERMS	FRACTIONS TERMS
5%		
10%		
20%		
25%		
30%		
40%		
50%		
60%		
70%		
75%		
80%		
90%		
100%		

LESSON 3: One Percent of What?

Add 1%, 2%, 3%, and 4% to the chart on page 129. Use the spaces above 5% for those percents. In the blank row after 10%, insert 15%. Complete the rows for each of these percents.

In what ways is finding 1% similar to finding 10% of a number? In what ways is it different?

LESSON 4: Taxes, Taxes, Taxes

If you read a study that said, "43% of biology majors have a high level of religious commitment," what are two ways you could determine 43% of 700 biology majors?

Midpoint Assessment: Meeting Your Goals

What do you know now that you did not know before starting the unit?

List two ways in which what you have learned will be helpful to you in your life.

How do you think you are doing with percents and fractions? What questions do you have?

LESSON 5: Fold and Figure

How can you prove that $12\frac{1}{2}\%$ equals $\frac{1}{8}$ of any whole? Show your reasoning with pictures, words, and numbers.

LESSON 6: Give Me a Third

How is $\frac{1}{3}$ different from 30%? Show your reasoning with pictures, words, or numbers.

LESSON 7: Order, Choose, and Change

In your opinion, what fractions and percents are most useful to know? Explain your reasoning with examples that include pictures, words, and numbers.

LESSON 8: Where's the Fat?

Investigate the foods at your home. Are they low in fat, medium in fat, or high in fat? If there is something else that you are concerned about (e.g., sugar, salt, or carbohydrates), then check the content of those in your foods. Give some examples.

CLOSING THE UNIT: Put It Together

Review work you have done in class and on your own. Pick out two assignments you think are your best work or show where you learned the most.

Make a cover sheet that includes

- Your name
- Date
- Names of the assignments
- A picture of a fraction, decimal, or percent of your choice. You can use numbers, number lines, or words with your picture.

For each assignment that you pick,

- Write a sentence or two describing the piece of work.
- Write a sentence or two explaining what skills were required to complete the work.
- Write a sentence or two explaining why you picked this piece of work.

> Take some time to look through your work.
> Try to remember each class you attended.
> What did you do?
> What did you learn?
> Look back at your *Reflections* and *Vocabulary* to get more ideas.